Orianna Fielding

# The Essential Digital Detox Plan

## How to achieve **balance** in a digital world

**FOR SASH**

THIS IS A CARLTON BOOK

Published in 2017 by Carlton Books Limited
20 Mortimer Street
London W1T 3JW

10 9 8 7 6 5 4 3 2 1


Original design concept by Orianna Fielding

The material in this book was first published in 2014 in *Unplugged: How to live mindfully in a digital world.*



A CIP catalogue record for this book is available from the British Library.

ISBN 978 1 78097 905 2

Printed in Dubai

# Contents

# Preface

**Connectivity now permeates every area of our lives. Technology has enabled us to do things that ten years ago were unthinkable. It has also completely changed many of the elements that historically formed the foundation of how we lived and interacted with each other, increasingly adversely affecting the most essential and fundamental elements of being human.**

My teenage years were lived in the pre-mobile, pre-Internet era. We had to call to arrange to plan to meet someone. We had to actually speak to someone to ask something in person or on the phone. We used "snail mail" for letters, cards and invitations. We did not have the world at our fingertips, but we were engaged in our daily lives because we were not distracted.

For me, digital technology in its early form was an "enabler". It provided a whole new series of life choices, made possible by being able to work from anywhere. It made it possible for me to move from the urban metropolis of London, where I grew up and spent the early part of my adult life, to the rural Mediterranean idyll of the coast of the Costa Brava in northern Spain, in my quest to find the ultimate in quality of life. In the early days, pre-smartphone and before Wi-Fi, I was able to be digitally connected via an airborne ADSL cable, enabling me to work from a studio nestled on a remote hilltop location, surrounded by a collage of lingering forests, underscored by a sea permanently merging with the sky like a pair of faded jeans. Perfect live/work balance? Tick.

Alongside the commissioned design projects I undertook for my clients, being immersed in such an unspoilt natural environment acted as a catalyst for me to search for a deeper meaning and purpose in both my life and my work. But it took time. In the early days I found myself physically present but mentally elsewhere. I was "here" but not really "here".

Really being present meant connecting with the seasons, learning the names of the different winds, recognizing the cycles of the moon and reading the ever-changing personality of the sea. Ultimately I learnt how to "be" by going to local farmers' markets. There I learnt to appreciate the beauty of imperfection, the splendour of a misshapen tomato, appreciating the real meaning of "slow". I had to learn a new rhythm, one without a preset time limit for everything. Where queuing for 20 minutes to buy some fruit was "just how it was" because everyone in the queue spoke to each other and wanted to share their stories. It was there, waiting to buy my imperfectly shaped, local, seasonal produce, that I began to really connect with where I was and learn to appreciate all the moments and experiences that really matter.

But my newly found connection with life was about to be seriously compromised with the advent of the smartphone. Suddenly I didn't need to be in my studio on my laptop to receive my emails and send projects through to clients as I could be mobile. I was free. I could access information wherever I went. It was a revelation and totally life changing. Although undeniably an incredible tool, I think I was slow to realize that having access to the world in the palm of my hand also meant the world had access to me. As my euphoria at being able to be connected no matter where I was wore off, it was replaced by a debilitating dependence on needing to be connected at all times.

And although I was *digitally* connected, I began to feel more and more *personally* disconnected from my surroundings as my virtual life was not giving me any real nourishment. It provided a lot of noise, butI could no longer find the inherent melody and rhythm of my daily life. Given that I was living in the sort of surroundings that are viewed as the ultimate "off-grid"environment – the sort of place where weekend "digital detoxing" retreats might take place – I realized that the problem could not only be viewed as relating to a purely urban demographic. It was a problem that was widespread, and one that required a solution that would work in any environment. The problem was personal and it stopped me in my tracks. I realized that what had at first been my lifeline had little by little started to strangle me. My digital dependence had become a habit filled with avoidance techniques and constant distractions that allowed me to avoid being with myself. I found that without realizing it my reliance on my digital devices had gone from expanding my life to controlling it.

I took a hard look at myself and found that despite being surrounded by an exquisite natural landscape, I was living a reductive, ultra-controlled digital life, where nothing was messy, chaotic or emotive. I realized I needed to step back from digital information long enough to be able to find a way to be "present" and come back to myself. I had to stop living in my head. My body was starved of attention and my other senses were suffering from abandonment anxiety.

So I decided to make a commitment to myself to find a way of re-establishing a digital/analogue balance in my life, so that I could once again control my digital world instead of it controlling me.

*Unplugged* is the result of the extensive research and personal experiences that fuelled my journey from London through the rural hills of the Costa Brava to the ultra-urban metropolis of New York. In my quest to discover the optimal method of creating a new digital protocol I have met and consulted with both inspiring professionals and game-changing individuals who have shared their unique take on treating digital overload. *Unplugged* is the distillation of what I have discovered on my journey to find a new digital road map that will enable us to re-establish a fully integrated online/offline life balance and reconnect with the essence of life.

Orianna Fielding

# Introduction

## Are you "on" 24/7?

**Technology and media are dominating our lives as never before. Increasingly, we are beginning to feel as if we are drowning under an electronic avalanche of incoming emails, texts and instant messaging. We are struggling with the effects of this continuous deluge of digital data and are becoming aware of the impact that being "on" 24/7 is having on every aspect of our lives.**

We are living in a culture dominated by digital excess. Many of us, even though we are already suffering from digital overload, seem to be permanently searching for more information, fuelled by an anxiety that we may be missing something – or worse, missing out on something. This "fear of missing out" has been labelled "FOMO".

However, that same technology that enables us to be connected digitally may actually be responsible for disconnecting us from our real lives. While we can feel empowered by our connection to a digital global community, very often that same digital connection removes us from experiencing real connections with people in our immediate environment. Often if we are engaged in "real life" offline activities, they are frequently interrupted by alerts from our digital devices, demanding our attention.

Studies have shown that digital over-connectivity can also be responsible for causing symptoms of depression and social anxiety, thanks to the lack of real human connection. While our digital life celebrates connectivity, without real meaning and connection our actual lives have no anchor, no core to sustain us.

Our "digital overload" has also manifested another side effect. It has enabled us to successfully avoid spending any time with ourselves. We have embraced the digital tools of connectivity, noise and activity and use them as a new way to live life virtually, instead of actually. Like a new version of the game *Second Life* for real people.

There really is such a thing as being too "plugged in" to our electronic gadgets. Increasingly, and backed up by substantive medical research coupled with the statistics showing rising incidences of IAD (Internet Addiction Disorder), this is clearly a growing issue and one that won't go away without recognizing it as a disorder that needs treating. Randi Zuckerberg, the former Facebook chief marketing officer, recently said herself that it was time for a "digital detox". While she views our smartphones as "an amazing tool" she qualifies this by saying that "we own our devices, they don't own us".

Overleaf is a list of the top 12 warning signs of digital overload. If you find that you identify with more than six of the signs, this may be an indicator that it might be time to undertake a digital detox and reboot your life.

**"Connection is inevitable. Distraction is a choice."**

Alex Soojung-Kim Pang

# 12 signs of digital overload

1 Checking your digital device first thing in the morning, getting up during the night to check for messages, and regularly using some form of digital device in bed

2 Slipping away from activities with people in order to check email or social networking sites

3 Checking your smartphone while at a meal with others

4 Bumping into someone because you were paying attention to your smartphone instead of looking where you were going

5 Spending little time outside, rarely taking breaks and often eating at your desk

6 Finding it hard to complete a task without frequently breaking away to check email or unrelated websites, often checking the same sites repeatedly within a short period of time

7 Getting distracted easily even when offline and finding it hard to focus fully on one area, or finding yourself unable to switch off your multitasking tendencies even when you're not multitasking

8 Spending little time in face-to-face interactions with friends. Choosing to spend time online rather than going out, preferring to use Skype or FaceTime to see people, even if you live near each other

9 Being with family members but spending most of your time at home in separate rooms interacting with screens, often finding that one monitor is no longer enough to suit your needs

10 Frequently using digital devices to entertain and keep a child occupied instead of talking, singing, playing or reading to them

11 Going online or using a digital device when you feel stressed or want to avoid an unpleasant task. Using the Internet as a way of escaping from problems

12 Wanting to stop using your smartphone and finding that you just can't, having tried repeatedly but unsuccessfully to control your Internet use

# 1 LIVE

## Are you living an "i life" or a real life?

**We are drowning under a digital flood of information. Much of the incoming information we receive is subscribed to or requested by us and the rest is generally unsolicited. However, as well as managing the information, most of the incoming content requires some form of interaction or response from us. We are spending increasing amounts of time online just trying to manage the 360-degree deluge of content. Our time, rather than being spent productively, is being ambushed by the checking, responding to and managing of both personal and professional emails, texts and social updates, leaving very little time for actual living.**

Our greatest attachment seems to have become to a digital device. Increasingly we are spending more time interacting with our smartphones than with anyone or anything else. The people we would like to spend time with are spending their time connecting with other people contained in a 2-inch screen.

Real-life conversations are becoming the exception, being replaced by the impersonal, managed communication of texts and emails, where emotion is expressed by graphic symbols such as emoticons rather than by the sound of the expressive inflections when we speak. Conversations are being replaced by reductive soundbites and as we are communicating less, is there a risk that our feelings will become abbreviated too?

As comedian Louis CK said in his now-infamous interview on *Late Night with Conan O'Brien*, which clearly tapped into the zeitgeist of public feeling as it subsequently went viral on YouTube: "You need to build an ability to just be yourself and not be doing something." He feels that this is what smartphones are taking away: our ability to just sit there. The message from him is clear. We just need to sit with ourselves, by ourselves, and "be". We need to stop trying to avoid feeling what we feel by creating digital distractions. That's being a person. It seems that his message hit home – over 7 million people viewed the YouTube video of his interview. (http://www.youtube.com/watch?v=5HbYScltf1c) As Jonathan Safron Foer said in his article for the *New York Times*, titled "How Not to Be Alone", "Technology celebrates connectedness, but encourages retreat."

The complex and unresolved love/hate relationship we have with technology comes from the fact that it is both an enabler and a disabler. It has enabled us to do things we couldn't have imagined possible just 10 years ago, yet it has also disabled many of the things we have done for centuries, adversely affecting many of the core fundamental elements of being human. We are engaged in a struggle to find a balance between our emotional intelligence and our social development in our attempts to keep up with the speed of new technological developments. The imbalance between our digital life and our real life is causing us to feel anxious, inadequate and, ironically, isolated. It seems to be symptomatic of a deeper disconnect. Although our digital connections on one level make us feel as if we are connected, in reality they are not nourishing us. It's like eating fast food, which delivers a lot of calories that temporarily make us feel full, but ultimately does not deliver any real nutrition. As human beings we want to experience emotion face-to-face, not through a digital filter. We want to touch and feel and connect in a real way.

Yet digital connection, while negatively impacting upon many aspects of our lives, has also instilled a new form of continuity for us. Having the contents of our personal digital world in the palm of our hand, wherever we are, wherever we go, offers us reassurance. Like a tortoise carrying its home on its back, we carry our social life, our work life, our photos, our notes, our video clips, our blogs and our posts with us at all times. So that ultimately, regardless of where we are physically in "real time", our digital life remains constantly with us, outside of our real world context, experience or location.

**"It's like eating fast food, which delivers a lot of calories that temporarily make us feel full, but ultimately does not deliver any real nutrition."**

## Are your attachments defined by a paperclip or a person?

We are living life out in the public domain, presenting an edited version of ourselves. We are buoyed up by having thousands of virtual friends with whom we are digitally connected, but how many real friends have we connected with recently, or had a face-to-face chat with, not using Snapchat, FaceTime or Skype? We are living with the constant struggle of staying in touch with our real family and our digital family – the online players such as Facebook, Twitter, Instagram, Snapchat, Tumblr, Vine and Pinterest that inform our days. In reality, studies have shown that our actual experience of social networks is that they are creating a new form of social isolation.

Most of us have a digital layer that overlays and informs our daily lives. Technology is also responsible for distancing and isolating many of us from our real time "real lives". It now seems to have become the default setting to see groups of people with blue-lit faces who are physically together in the same place but who are all looking at their smartphones instead of connecting with each other. The loss of true connection and communication through its replacement with social networking, texting and other forms of digital communication is making us less mindful, less genuinely connected and less authentic. To be able to share our authentic selves requires more than the sharing of digital images and messaging. To build real relationships we need to understand who we really are. It is a process, and one that takes time.

Problems arise when the balance of our communication time shifts to the digital side rather than the real, face-to-face side; when we are actively choosing to replace opportunities for real-time live communication with digital messaging.

### Remember the original meaning of "like" and "share"?

What was the tipping point, when the technology we created to streamline our lives and use as a communication tool ended up taking over? When did the need to play out every aspect of our daily lives publicly to all of our virtual friends and followers across multiple social platforms become the norm? What happened to

PSYCHOBUBBLE

## Less, better

*One of my clients was talking about the loneliness and isolation she felt, despite having a husband, children, parents and hundreds of friends on Facebook.*

*"I remember when I was in my teens and twenties," she said, "I talked with my friends about everything, I felt really close and supported. Now, everyone seems to deal with the difficult stuff alone, then tell you a sanitized version after it's happened."*

*I shared with her my view that what seems to have become the contemporary ideal of "A Hundred Friends" on Facebook is actually preventing real intimacy. We are putting out carefully edited versions of ourselves and our lives which we feel others will approve of. And so a whole plethora of real, messy, complicated human emotions and experiences are being locked away into a box inside us labelled "not good enough for public consumption".*

*This leads to what I call "Shame, Secrecy, Isolation". A psychological trap where you feel that certain emotions – for example, anger, failure, envy, despair, struggle – are unacceptable, and so keep them secret. This then escalates until it feels like something shameful that can be revealed to no one. As the power of the shameful secret grows, so does the sense of isolation. And on it goes, in a vicious triangle. The best antidote to my client's feelings of isolation and loneliness is: "Less, Better". Can she build up trust, sharing, honesty and true intimacy with just one or two friends, in real time and space?*

*Try to meet up, face-to-face, once a week or more and bring your whole self and honest experience to that meeting. Leave your polished, edited, digitally-acceptable self at home if you want to feel fully connected, understood and accepted. Which is what, after all, most of us crave.*

**Jacqui Marson**

Chartered psychologist and author of *The Curse of Lovely*

living a private life of personal moments shared with friends, family and loved ones on a "need to know" or "really want to share" basis? At what point did our real-life friendships begin to take a back seat to our online social media friends?

When did we stop getting excited about meeting up with our friends because we had lots to tell them? Nowadays they already know everything that has happened to us at all times because it's already out there in the public domain and we've posted it online as soon as it happens! Why do we need to play out our lives on a public stage to a virtual voyeuristic audience by providing an open platform for them to randomly dip in and out of the sound and vision bites of our lives?

The early part of my teenage years were lived in a pre-mobile, pre-Internet age, and (inconceivably for anyone born after 1990) life without a mobile phone was fine. Everything happened exactly as it should. We did not have the world at our fingertips, but we were engaged in our daily lives. We were not distracted, trying to do everything, know everything and be everything at the same time. We did not have a public platform on which to play out our lives. The concept of a private life still existed. It was the norm unless you were really famous.

Now, instead of treasuring having a private life, we seem to be playing out our most personal moments in a global digital public arena. Our perspective of life is clearly changing. We no longer seem capable of just seeing and experiencing "the moment". Just looking at something and appreciating it is no longer enough. We have a compulsive need to document every moment and experience and instantly share it with our digital disciples. We now look at everything we experience, and everyone we meet, as something that requires our commentary or as a photo opportunity, to be snapped, uploaded and posted into cyberspace for the mass consumption of our digital followers.

According to University of California neuroscientist Loren Frank, the reason that we can become so easily addicted to the constant digital attention we get when someone "likes" us on Facebook or Instagram, or tweets us, or we receive an email, is that the attention gives us a "nudge". He says: "It's a small jolt, positive feedback that you're an important person," and that need for digital validation can quickly become quite addictive. The digital attention we receive can bolster our self-esteem but it also can have a negative impact on how we feel, particularly when the line between our digital popularity and our real-life popularity becomes blurred.

## How social are we?

Despite our growing awareness of the time we spend online, and how that impacts on our offline relationships, most of us use some form of social media and are connected digitally over several platforms. We all voluntarily form part of an immense global online community and each of us have our own reasons for using social media, but the reasons why we use social media can be very different.

Every social media platform has developed its own unique identity, initially by design, to appeal to its target market. But as the statistics show, social media is no longer the preserve of the young or the technically minded. In fact Facebook has recently seen a substantial shift in its demographic with an increase in the number of older people

## The top 7 reasons why we use social media sites

1 We may be in need of company and find it in our online communities.

2 We may be looking to extend our "real world" communities online, enabling further communication as a support to our offline activities.

3 We may want to use social media as a tool to find and reconnect with people we have lost touch with.

4 We may be living in an isolated location or just feel isolated and turn to social media for an alternative form of interaction, using it as another way to build friendships online.

5 Many of us can often feel that we don't have a voice or find it difficult to vocalize our thoughts and feeling face-to-face. Social media gives us all a voice, instantly. We have a forum to share whatever thoughts we have, knowing we will be heard.

6 Ironically, social media platforms provide us with the ability to control how much of us we want to reveal and to whom. Through the use of privacy settings we can be the guardians of our own experiences, ultimately choosing with whom and when to share them.

7 We all want to feel as if we belong and, to a degree, becoming part of an online community, identifying our tribes and joining those groups gives us a sense of being a member of a club where we matter.

starting to use the platform, which in turn is causing its younger users to switch away from Facebook to newer platforms such as Instagram and Snapchat.

**"Care about what other people think and you will always be their prisoner."**

Lao Tzu

Most of us interact with social media in some form and, as shown, we have our own personal reasons for doing so. However, the ways in which we use social media sites also vary enormously. Some of us love to create content, while others prefer to be passive participators and mostly browse through other people's content, occasionally posting a comment or photo. Many people love to use social media as a forum to share their lives, while others use it for business. Each of us has our own unique way of using social media and our social media behaviours have been categorized into different types.

Based on the findings of a study conducted by consumer research data company Aimia, analyzing the way in which social media networking sites are used in the US, six different types of social media users have emerged. I have categorized them over leaf.

## Is social media making us depressed?

A study by the University of Michigan, observing 82 Facebook users over a 14-day period, found a direct correlation between the time they spent on Facebook and the negative impact upon their self-esteem. The study showed that as the time the group spent on Facebook increased, their feelings of depression also increased and their feelings of wellbeing decreased. The project's lead researcher, Ethan Kross, explained: "On the surface, Facebook provides an invaluable resource for fulfilling such needs by allowing people to instantly connect." But instead of increasing the feeling of wellbeing, these findings suggest that spending a lot of time on Facebook may actually undermine it. According to Kross, the study findings went on to show that "the more time someone spends on Facebook, the worse their mood outcome". These findings were further backed up by a study by the American Academy of Pediatrics which also found that both children and teenagers can develop what they term as "Facebook depression" as they feel overwhelmed by all the positive status updates and happy photos they see their friends posting. London psychotherapist Jacqueline Palmer, who specializes in individual counselling and psychotherapy and provides help and guidance to cope with feelings of isolation and being overwhelmed, summarized this syndrome, calling it "compare and despair".

As we become more aware of the negative impact of our dependence on social media, we may already be seeing the start of a backlash. Even Millennials are now starting to re-evaluate their social media activity. Increasingly Facebook, Twitter and Instagram accounts are being either temporarily disabled or deleted.

## Which type do you think you are?

- **The Motivator**

**Motivators are the drivers of most social media networks.** They create content and are actively engaged with their social media platforms and use them as a forum for self-expression. Online privacy is important to them and they address this by monitoring their online conversations carefully. Motivators pride themselves on having the most accessible social networks of all the six categories. They often interact with brands of their choice and make willing brand ambassadors for their selected brands.

- **The Concealer**

**Concealers are mostly hidden from view and log on to social media sites infrequently**, maybe no more than once a month. Generally they tend to be male and in the 65-plus age range. They have a general distrust of social media sites and mostly have little interest in publicly sharing their daily activities or interests with anyone, particularly people they don't know personally.

- **The Passenger**

**Passengers are generally passive users of a single social media platform.** They may, for example, feel under pressure not to be left behind and excluded, so they reluctantly join Facebook. Passengers mainly use social media to have some form of online relationship, but do not actively post or drive conversations.

- **The Spectator**

**Spectators may hover between several social media networks, more as observers than active participators.** Rarely posting or interacting, they use social media mainly to keep up to date with the online lives of others inside their social networks, but feel hesitant about sharing personal information or details of their own lives. Spectators, although mostly passive bystanders, still want complete control of their online presence.

- **The Insider**

**Insiders are very proactive users of essentially one social media network platform**, who congregate mostly on Facebook. They are generally women, and they frequently share photos, provide regular status updates and post comments. They are very active online and have influence within their personal network of family and close friends.

- **The Integrator**

**Integrators have a presence on several social media sites.** They participate actively on all of them and keep up with daily posts and news across the various sites. They like to follow brands, particularly in order to receive special promotional offers. Although they are active on several social media platforms they also value and have an understanding of the importance of controlling the privacy of the data they share. Many of their offline friendships are first formed online, and they have great influence within their networks.

Which type or types do you identify with the most? Most of us move across several types of social media personas, which will continue to change as we do. With age, and our evolving interests, coupled with changes in our personal lives, the way we use social media is not a constant but rather a shape-shifting, evolving journey.

This has been highlighted by a rash of high-profile names such as *Divergent* star Shailene Woodley and rapper Nicki Minaj deleting their social media accounts. Woodley explained in an article for *Marie Claire* that although she had 20,000 followers she found the narcissism attached to using social media, posting and tweeting every aspect of her life, "disgusting". She deleted all her photos on Instagram essentially because she was tired of living in a culture of "me".

**"Attention is the rarest and purest form of generosity."**

Simone Weil

There is even a National Day of Unplugging (NDU) which takes place annually on March 6–7, spanning a 24-hour period from sunset to sunset. The event has a supporting website that enables anyone who wants to "unplug" to upload a photo of themselves with a sign saying "I Unplug to...", which users can personalize with their own reasons for stepping away from social media for a day. (http://nationaldayofunplugging.com)

This annual unplugging project was created by the creative network Reboot. The project is an extension of Reboot's "The Sabbath Manifesto", a contemporary version of the biblical tradition of reserving one day of the week to take time to relax, reflect, unplug, rediscover nature, and connect with loved ones.

Unplugging has even been adopted by global fashion brand Giorgio Armani's fragrance division to raise funds for UNICEF's Tap Project. They will donate 25 cents to UNICEF's US fund for every minute a mobile phone user goes without his/her mobile phone on the UNICEF Tap Project mobile web application, over a four-week period, and are hoping to raise a significant amount of money to support UNICEF's water and sanitation programmes.

## Our disconnected lives

The underlying anxiety and loneliness that being disconnected on a human level makes us feel seems to be reaching epidemic proportions. As human beings, we are missing the subtleties that face-to-face interaction enables us to experience.

I consulted with Howard Cooper, a Rapid Change therapist. He creates personalized individual treatment programmes that combine hypnotherapy, neuro-linguistic programming and Thought Field Therapy to treat a variety of disorders such as the anxiety caused by digital addiction and the fear of missing out (FOMO). His technique focuses on "re-booting our relationship with our digital connectivity and finding a new way to manage our online and offline relationships". He says, "We are living in a world that supports the overuse of technology and is overriding our experience of the 'real world'." His solutions look at short-term goals to lead to long-term changes.

Cooper accepts that there is a sense of safety in feeling connected digitally, but believes this has a counterpoint: the loss of the sense of real, genuine, human connection. Through his personalized technique we can learn "how to become more by connecting less" and concentrate on how we can begin to trust our own real-time, unedited and random experience of life outside of the edited, managed and controlled "director's cut" digital version of our life that we present online. The issue of "attention deficiency" within our human connections led him to share the story of a childhood experience that had a profound impact upon him.

Cooper explained that at school he would occasionally be required to go to his Deputy Headmaster's office to get advice.

> "I would knock on the door and his loud booming voice would shout, 'Come in'. So I opened the door and there sitting behind a large desk in a whirlwind of activity was the deputy headmaster. As I entered the room, he would look up and see me and say, with a beaming smile and a twinkle in his eye, 'Have a seat, I will be with you in a moment!'
>
> At this point he did something so powerful that I have remembered it to this day. He proceeded to turn his computer screen off, he piled up all the papers on his desk and put them to one side. He then took the phone off the hook, pushed the keyboard and mouse away and closed the open book he had been looking at. He then stopped, stood up and picked up his chair and moved it around the desk so he was sitting directly in front of me. He then took a deep breath and looked straight into my eyes as he sat down, and said, 'I'm all yours... what can I do for you?'
>
> In fact, whenever I went to see him, he followed the same rituals, and the effect it had on me, and the other people who went to his office, was profound. For at that moment, I felt like the most important person alive. For a brief moment in time, the busiest man in the school would stop everything for me, and it never ceased to make me feel special. In fact, each time he put his items away it made me realize that he was preparing to be totally present with me, totally and completely connected to our interaction."

**"The greatest gift you can give someone is your time because that is something you can never get back."**

Now, this was someone who really understood the importance of giving our complete attention to whoever we are with, to be able to fully connect with them.

# Techno babies

## The birth of a digital nation

Digital technology has enabled us to share our most life-changing moments, including the most private of all: giving birth. For many years Skype has been the established medium for connecting family and friends long-distance, and now it has also become the preferred platform for sharing the birth of a baby long-distance with our partners. We have come a long way since the time when it was the exception for fathers to be present for the birth of their own children; today, via digital technology, they can participate remotely wherever they are. Birthing videos have been around for a long time and they have now become elevated to an art form, with smartphones being strapped to tripods and time-lapse photo apps being used to create birth movies for parents to keep for posterity. Social media platforms such as Facebook have also provided a forum for the sharing of "birthing progress reports" with friends and family via continuous status updates, with gaps in between contractions being used to read comments offering support and encouragement, a new form of virtual group birthing.

The use of technology in parenting offers digital solutions to practical issues, such as baby monitoring, baby entertainment with musical sleep-inducing playlists and bedtime stories. The increased uptake of these forms of "electronic babysitting" reached a point of no return with the launch by a major globally recognized children's brand of their new Apptivity Seat, marketed as a new activity centre for newborns and toddlers. Nothing unusual there, except for the add-on arm that extends over the seat and which has a flat iPad case attached to the end of it, designed to protect the iPad from spills, dribble and fingerprints. To vindicate themselves after the international outcry from parents that followed its launch, the brand explained that when an iPad isn't in the case, there is a mirror there to reflect the baby's image. They also explained that the arm can be moved and repositioned away from the baby so that parents can actually interact with their babies face-to-face. The American Academy of Pediatrics' recommendations are that screen time for babies under two years old should be limited.

According to Dr Troseth, a psychologist at Peabody College at Vanderbilt University, infants and toddlers, unlike school-age children, have no idea what's going on when they are watching a video or streamed programme, no matter how artfully or creatively the content has been crafted.

We are still amazed by the speed with which babies who are barely able to grasp a coloured crayon are able to work out how to use a touchscreen. The report from the American Academy of Pediatrics found that for every hour a child under two spends in front of a screen, nearly an hour is lost interacting with a parent or carer, and this subsequently reduces their explorative and creative playtime by 10%. It recommends that parents look at setting "media limits" for babies and toddlers. For many parents and professionals the idea that babies could be exposed to digital media is complicated enough to assimilate, so the further requirement to set a screen time limit for babies seems inconceivable to many.

Our digital devices can undoubtedly be viewed as a pioneering educational tool, containing several terabytes of educational facts within their deceptively slimline exterior. Theoretically one can see the potential educational benefits that digital devices such as iPads and other similar devices could offer to a toddler's receptive and absorbent

## Privacy: the digital dilemma

An interesting dilemma has arisen through our need to post and share personal pictures of our children with our family and friends on social media, aside from the ongoing discussion on privacy. This is an issue that needs addressing as it affects the long-term digital presence of our children.

The question of establishing our children's online presence prior to them being able to decide whether they are OK with the content their parents are posting is a serious topic with long-term irreversible consequences for our children. Sharing our lives through pictures has become a normal part of our daily activity and, globally, picture-posting statistics have reached extraordinary new levels.

According to statistics compiled by www.mediabistro.com, in 2013 a total of 240 billion photos had already been uploaded to Facebook. Every day a further 300 million pictures are uploaded to Facebook via Instagram, and the total number of Instagram users (which is now owned by Facebook) has risen to 90 million.

While the innocent baby pictures posted by parents enable extended family members and friends to participate and share in the lives of their loved ones, they are also establishing a permanent digital footprint that will remain in place regardless of whether the children would have chosen that same content to be posted about them or not. Once they are old enough to decide, the option to rescind or delete posts will probably no longer be available to them as they will be in the public domain, forever floating in cyberspace. This issue has led many parents to start removing previously posted pictures of their children from social media sites as awareness of the long-term consequences of sharing baby pictures comes into sharp focus.

brain. Recent statistics show that 77% of parents actually believe using a tablet is beneficial to their child. The question is, who benefits the most from putting an iPad in front of a baby or toddler? In terms of managing parenting time I suspect it is the parents who benefit from the hands-free parenting that this new form of "electronic babysitter" or "iPad Nanny" provides.

Another consequence of the dependence of an increasing number of parents on digital devices is that they now also have the ability to digitally monitor their babies and toddlers 24/7, tracking their every movement, sound, breath, and keeping an eye on their general health. A little like being tagged and under house arrest, for toddlers in particular, being continuously monitored and then presumably over-protected from the possibility of real life getting in the way could, instead of protecting them, actually prohibit natural play and learning.

## Out of the mouths of babes...

Interestingly, children themselves are becoming much more aware of the wider consequences of having an online presence. In a recently commissioned video Mozilla shared some children's concerns about their future and the type of web that they want to inhabit. The company had children read out a statement expressing concern over the way the Internet is used, and the need for adults to ensure they are passing on a free, safe and open web to future generations. Lines from the statement include 'My identity is mine', 'I need technology I can trust' and 'Remember to put me first!'. (http://www.upworthy.com/why-are-her-baby-pictures-under-surveillance?c=reccon1)

On a lighter note, our obsession with our digital devices has led to a rash of new "tech-inspired" baby names such as Siri, Mac and Apple – with one of the most literal being a girl called Hashtag, according to the annual report from parenting advice website BabyCenter.

According to the findings of a national survey undertaken by Common Sense Media on the use of mobile devices by children in America, 17% of children under the age of eight use digital devices daily. They found that the use of digital devices in families with children under eight years old has increased dramatically over the past two years, with 75% of young children having access to some form of mobile device and 38% of babies and children under two years old having used a digital device.

CEO of Common Sense Media, James Steyer, commented on the findings of the survey by saying that the mobility of our digital devices has caused a major shift in how such devices are used. Whereas in the pre-mobile era parents could measure and control exactly the amount of time and how their children were engaging with screens, monitoring and controlling toddlers' screen time has now become more complicated as devices can now accompany their children wherever they go.

The amount of digital media that toddlers and young infants consume can have a long-term negative impact on their ability to learn, on their behaviour and ultimately on their ability to acquire social skills. Parents who feel vindicated by downloading educational apps onto their infants' mobile devices are still recommended to monitor their children's screen time closely. The type of interaction a toddler will have with a mobile device or iPad, even if it is loaded with educational apps, cannot provide the same form of learning and personal development that interacting in the real world provides.

Babies and toddlers need real-world stimulation. Despite the fact that this could be viewed as very "last century", according to the American Academy of Pediatrics screen time doesn't provide the same education as that offered by the real-world experiences that a toddler actually needs. Colour recognition and motor skills may be developed through interactive computer games where they may learn to move objects, match colours and pair shapes, but ultimately while they are using these limited skills to interact with a screen, toddlers are missing the valuable experience of recognizing those forms and colours in the real world, something that is essential for their development. The same applies to their language development. According to Kathryn Hirsh-Pasek, professor of psychology at Temple University, the more a child hears real conversation, with eye contact and inflection and sentiment, the more their comprehension, language skills and vocabulary will increase.

## Scream time

Addiction to digital devices, in particular the iPad, already affects a large percentage of young children. One in three children has used an iPad before they can even talk, and two-thirds of children are using one on a regular basis by the age of seven. The seamless adoption of the touchscreen "swipe" by young babies was perfectly exemplified in a video called "A magazine is an iPad that doesn't work", showing a real-life clip of a one-year-old baby, who has clearly grown up amongst touch screens, trying to unsuccessfully swipe and select magazine pages. The video went viral and has had to date almost 4.5 million views.

iTunes, in order to feed the market for infant-targeted content, has launched more than 40,000 children's games and apps. The combination of this vast selection of easily accessible and compelling digital games and apps has led to an increasing number of children, as young as three, demonstrating signs of "IAD" – "Internet Addiction Disorder".

Dr Richard Graham, consultant child and adolescent psychiatrist at the Tavistock Clinic, runs Britain's first rehabilitation centre for technology addiction for children and young adults. Dr Graham commented that when the child's digital device begins to have more influence over their behaviour than anyone or anything else, that is the pivotal moment when parents should be aware that there really is an issue that needs to be addressed. In order to assist parents in evaluating whether their child may be developing a technology addiction, he has published a checklist that focuses on five key areas.

Parents need to seriously consider the impact that the overuse of technology is having on their children. Children take their cue and learn by copying the adult behaviour around them, which mostly centres on observing their parents' use of digital devices in their home environment. To young children, the continuous use of smartphones and iPads, coupled with the level of attention their parents or caregivers often give them, can send out a signal that being permanently attached to their digital devices is both desirable and acceptable.

Parents are recommended to "walk the talk" and educate by example, by ensuring that their own online activities reflect the positive relationship with digital devices that they are encouraging their children to have. Continuous or excessive use of the iPad or other digital device can isolate a child and encourage anti-social behaviour by taking them away from family activities. Instead, parents can instigate shared screen time, where they sit with their children and interact with them while watching something together.

Ultimately the simplest solution is to prevent the problem before it starts. Parents and care givers can teach their children by setting an example. They can determine the way their young children spend their time, particularly during those early formative years. The need for children to go out and play has not changed. However, according to the Royal Society for the Protection of Birds, four out of five children in the UK are not connected to nature.

Like previous generations, children need to be allowed to get cuts and bruises from playing outside in the fresh air, connecting with the real world, using all their senses. Measured iPad use needs to be integrated with offline activities to form a balanced diet that combines physical activity and play with non-electronic activities. This will provide a counterpoint to their absorption of virtual experiences and teach them the skills they need to prepare for the real world.

## Bye bye baby

The type of anxiety parents would have had in the past with teenage babysitters mostly centred on their babysitter consuming the contents of the fridge or inviting their boyfriend over while they were out. Today the issues have changed, with the potential for online activity like Twitter and Facebook to distract them and even potentially compromise the safety of the children they are looking after. The relationship teenage and young adult babysitters have with their smartphones is different from ours. They grew up with mobile devices; they are part of their everyday life, which means becoming distracted while watching our children is even easier.

Facebook has over 14 million users between 13 and 17 years of age. If your babysitter falls into that category, texting while watching your child may not be your greatest concern, but the idea that your babysitter texted all night and ignored your kids is

**5 key signs that your child may be developing a technology addiction**

1. Demonstrating a general lack of interest in participating in other activities
2. Continuously talking about digital devices
3. Frequent mood swings depending on permitted time on digital devices
4. Withdrawal symptoms when their digital devices are taken away
5. Using devious behaviour to obtain screen time

disturbing. Of further concern to parents is that a babysitter could innocently post photos of their children, or check in on Facebook while out with the children, therefore sharing the exact location of where they are. In turn this announces publicly that your home could be empty. There have been incidences of burglaries taking place thanks to criminals keeping an eye on social networks for people posting about going on holiday or going out for the day.

So how do we deal with a babysitter or childminder's constant online activity while watching our children? There is a simple solution: communication. Talk to your babysitter and go through ground rules, explaining why these rules are important to you. This is the time to say it like it is. Be clear. If their online social media activity doesn't feel right to you, let them know. If you don't want your babysitter texting or tweeting while looking after your children, tell them explicitly. Be specific and emphasize that the most important thing is that your kids are safe while under their care, and explain why being distracted could have serious and damaging consequences.

**Some suggested guidelines to share with your babysitter while your children are in their care include:**

- **No social media**. Adrienne Kallweit, president of Seeking Sitters and author of *S.T.Y.L.E. – A Complete Guide for Babysitting Success*, recommends that babysitters do not post on Facebook, Twitter, Instagram or check in on Foursquare while babysitting. She explains that, "Some camera phones and smartphone applications have the ability to pinpoint the exact location of a photo or social media post, by clicking on the posted photo or status update."
- **No photos**. Make it a rule that your babysitter should never post or tag pictures of your child, even when they are not actually with them. "Tagged photos share the name and exact location of your children with everyone on the Internet," Kallweit points out.
- **No online "Babysitting" news flashes.** Posting social media status updates on when or where your babysitter will be alone with your child may be dangerous both for the safety of your children and for your babysitter.
- **Screen time**. If you have strong preferences about the amount and type of screen time your child consumes in your absence, you need to communicate that clearly by setting limits both with your children and the babysitter.

## Infinity and beyond

The World Wide Web has come of age. It grew up fast and without any boundaries, thanks to our awestruck response to the wonders that it offered us. We were so blinded by its potential that we didn't notice our relationship with it was beginning to spiral out of control. For Millenials it is hard to imagine a world without the Internet. We are now just beginning to realize that beyond the amazing and powerful tool is the potential for it to dominate and control every aspect of our lives. As a result we are now starting to examine our relationship with it and are searching for new "human operating systems" that will enable us to restore the balance between our online and offline worlds so that we can reconnect with our "real analogue life" in real time. Our online today can inadvertently become our indelibly marked tomorrow, forever floating in cyberspace.

Teenage years are a time of exploration, self-discovery, trying new things and making mistakes, none of which is new, apart from the fact that now the road map to our personal development has a GPS and our every move is digitally documented in some form. In the past, if our experiments misfired, a few trusted close friends would share the embarrassment, support us through it and we'd all move on. Now those achingly embarrassing moments that are a normal part of every teenager's life have the power to turn up to haunt them at a very inconvenient and inappropriate point in the future. The irony of the permanence of the digitally uploaded life is that the most personal moments are offered up voluntarily and shared for digital posterity.

Those shared photos and texts that seem such a great idea at one particular moment in time all have an infinite digital shelf life. Once posted, never forgotten. Perhaps that explains the huge popularity of Snapchat, a mobile messaging app that addresses the issue of moments in time being captured forever by sending images that disappear after a few seconds. However it seems that even on Snapchat the security of your content cannot be guaranteed, with their recent announcement of additional new features such as text messaging and chatting, and a Skype/FaceTime-style video conferencing capability. Snapchat's USP was its disappearing messages, but with these new additional features it transpires that apparently anyone will be able to just tap on a text message to save it without the sender ever knowing. Even more surprisingly, if you and your friend are chatting at the same time, the thread of that chat does not disappear at all, and neither do any of the pictures you've put into that thread.

## Mission control

I think most of us feel under pressure to get through our ever-expanding, self-imposed live/work "to do" lists. We are always searching for ways to get more and be more, in our quest for perfection. Our Nirvana seems to be a mythical perfect life, over which we have 100% control so that we can avoid the messy, awkward and painful parts. I think that we believe that our digital lives enable us to achieve that. We view our online life as one that can be perfectly controlled via a full digital menu of "life erasing" options. We can edit the public face we present to our digital followers and we can embellish content to create the impression that we are living the life we want to have, fully controlling how we want to be perceived. But in reality life is not perfect, and none of us is perfect. Our flaws are actually what make us human, unique and relatable. We can't control life. Life IS imperfect and changeable.

We can spend as much time editing, Photoshopping and manipulating our online personas as we want. However, in the cold light of our analogue day, the harsh reality is that it is the unedited version of ourselves that looks back at us from the mirror. The challenge now is to find a way to reconcile our various digital selves with our real-life offline selves, each with their own often conflicting identities. It is important to remember that living parallel realities that layer our digital lives over our analogue ones can have real-life consequences.

Many of us, from the pre-Internet generation of baby boomers to the online generation of Millenials, are still working out how to navigate our way around our digital world. We are all still trying to work out how to be, in the contrasting contexts of our worlds.

Teenagers in particular feel a lot of pressure to have a popular online persona, qualified by the amount of likes, retweets and general attention they receive. They are living with the equivalent of a daily digital barometer that determines their popularity and in turn bolsters or destroys their confidence.

Danah Boyd MA, Ph.D and senior researcher with Microsoft, explores this topic in her book *It's Complicated*, in which she looks at why teenagers seem to behave so strangely online. She explains that they are trying to work out how to behave in their digital world and learning how to navigate their online interaction with their friends. It is a complex path that is continuously shifting as the parameters change, reflecting digital trends, new technologies and the fact that they are growing up.

For many teenagers their online world gives them an opportunity to have a social life that perhaps offline would not be possible. For them, their smartphone is the lifeline to a "happy place" that they can escape to, without the restrictions that school and parental interference would normally impose.

The yin to the yang of teenage escapism online is that as a teenager or young adult it is hard to take a long-term view of the consequences of an Instagram picture of a group of friends at a party, or an artfully-shot selfie on Facebook. Teenagers live for the moment but those moments, both good and bad, are digitally documented for all to see, share and refer to in later life. In many cases, content innocently posted by teenagers online has come back to impact on their adult life, negatively affecting job applications, relationships and reputations. The essential thing to remember is that once you've uploaded an image online, you have lost control of that image for ever.

## Modern family

Electronic devices have irrevocably changed the foundation of modern family life. The home which had always served as a shared haven, a private escape from the intrusions of our public life outside, has mutated instead into a "media hub" for its permanently plugged-in family members. Parents and their children, although sharing the same physical space, are disconnected from each other by their permanent connection to digital devices that replace the time they would have previously spent together. The disconnect appears most

PSYCHOBUBBLE

## Connect with yourself

*In my book,* The Curse of Lovely: How to break free from the demands of others and learn to say No, *I suggest several tools and techniques that help us step away from virtual reality and fully connect with the person you really are. One of the most popular of these is something I call the "Three Good Things a Day" diary. Research has shown that constantly comparing ourselves to what we think are others' lives and achievements, as displayed on their social media sites, leads to a lowering of mood, feelings of failure and low self-esteem.*

*This exercise helps build confidence and self-esteem in ways that are NOT based on how many people "like" your post or picture. The Positive Psychology movement argues that if you keep this diary for 30 days it will help create new neural pathways in your brain that encode the habit of thinking positively about yourself and your qualities.*

*First, buy a beautiful notebook that celebrates your intention to do something nurturing for yourself. Then, divide each page in half. In the left-hand column, write down at least one thing each day that has made you feel good about yourself, that you are proud of in some way.*

*Then, in the right-hand column, write the personal qualities that this shows you have. This is not as hard as it sounds when you realize you can include all kinds of actions and tasks that you might otherwise dismiss: calling a friend in need (kind, caring, thoughtful), cooking a new recipe (creative, daring, experimental) or queuing for a much-wanted item (determined, focused, patient). This is about praising yourself, and not waiting for the praise and approval of others. Very empowering and no new technology required!*

**Jacqui Marson**

Chartered psychologist and author of *The Curse of Lovely*

noticeably in families with teenagers. Parents with children who are transitioning to young adulthood often feel separated from their children by a vast, almost unfathomable divide as they unsuccessfully compete for time and attention against their kids' online social family.

**"Asking a teenager to shut down their computer or look up from their smartphone has the same response as asking an alcoholic to put down their drink."**

According to a 2010 study of children aged eight to 18 conducted by the Kaiser Family Foundation, teenagers today spend nearly eight hours a day immersed in their digital worlds. The smartphone has facilitated continuous connectivity for more than 75% of all teens, according to a study conducted by Pew Internet and American Life Project in 2011.

Celebrated filmmaker Beeban Kidron's documentary *In Real Life* is a touching and thought-provoking film that explores the enticing, alienating and addictive nature of the digital world that teenagers and young adults inhabit. (http://inreallifefilm.com)

The film explores the darker side of the intense social pressures that teenagers experience within their digital worlds. It delves into their extreme anxieties and insecurities about image, in both their online and offline worlds, fuelled by unrealistic digital manipulation and comparison. Featuring a group of "talking heads", these teenagers openly discuss all the issues of growing up in a digitized world, from the desolation of feeling left out to the new forms of digital dangers such as cyber bullying, naked selfies and uncensored porn, and how this all negatively impacts on their ability to form healthy offline relationships.

Some of the most hard-hitting and disturbing parts of the film make very uncomfortable viewing, showing how the kidnapping of a smartphone can be used as a leveraging tool by young boys to get young girls to do almost anything to get their phones back.

According to Kidron, there is no question that Internet addiction exists and, as her film so clearly shows, it has a very pernicious side, especially for teenagers and young adults.

On their journey to becoming an adult, teenagers need positive role models to imitate. Instead, often their most immediate role models – their parents – are mostly too busy holding their smartphones to be able to provide a healthy example of a balanced, controlled approach to their own digital over-connectivity.

**"People may doubt what you say, but they'll believe what you do."**

Lewis Cass

# 8 positive steps for families to reconnect with each other

**The American Academy of Pediatrics has a series of practical recommendations to enable families to reconnect with each other:**

1. As parents we have to watch our own behaviour and lead by example, making sure we don't set screen time guidelines for our family while remaining digitally over-connected ourselves.
2. Be present when with family members. Listen. Give whoever is speaking your full attention, without holding a digital device in your hand and skimming emails at the same time.
3. Make mealtimes a device-free family time with a no-device rule during meals, to encourage real conversation, re-enforcing a "no meals in front of the computer" rule.
4. Bedrooms should also be a device-free zone in order to encourage restful sleep.
5. Monitor whether your teenager's real-life friendships are taking a back seat to their online social network friends.
6. Set time limits for screen time, making sure there is enough time every day for other important activities, such as schoolwork, hobbies that don't require a screen and spending offline time with friends.
7. Share tasks. Create a weekly shared roster of family tasks for each member of the family to complete.
8. Go outside. Find an activity that the whole family can participate in or even find a local team they could join.

## Driven to distraction

Even common sense seems to have "left the building" in terms of our ability to think rationally about our need to be permanently digitally connected. A key example is that we now seem to need laws telling us that we actually are not allowed to text or tweet while driving, because the urge to be connected at all times can overpower our normal thinking process.

In the US they have now had to introduce "texting stops" on the highway, in order to combat the epidemic of text-related driving accidents. These are dedicated places where drivers can pull over, stop the car and answer that text before they reach their destination.

Even car manufacturers have now responded to our need to be connected even while driving, with Ford for example launching a new Fiesta model that syncs a driver's smartphone with an audio text reader that reads texts aloud while they are driving. A host of other car manufacturers are now following suit. It seems incredible that we now need digital technology solutions to enable us to control our use of digital technology.

In May 2014, New York's Mayor Bill de Blasio launched the "Vision Zero" scheme, a crackdown on drivers talking or texting using hand-held devices while driving in an attempt to dramatically reduce traffic fatalities.

In order to drive the message home, New York's streets and highways are lined with signs saying "Zero distraction".

According to the NYPD, texting or talking on mobile digital devices is an ever-increasing problem despite the obvious dangers of distracted driving and has been shown to be responsible for a significant percentage of road traffic accidents, injuries and fatalities. Studies have also shown that, surprisingly, walking while texting could potentially result in more injuries than driving while texting.

## Out of con*txt*

A new trend has infiltrated the hallowed walls of our education system, namely the increase in use of texting and tweeting language, or "text speak", appearing in written school work.Children as young as eight years old are using the abbreviated forms of spelling learned from texting, and are applying them to their academic school work. This reductive approach both to language and writing skills can over time lead to a permanent replacement of formal established language rules.

The UK-based charity for learning disabilities, Mencap, conducted a survey in 2013 on the use of text speak in schools. Their findings, based on a sample group of 500 British teachers and parents, found that 66% of teachers confirmed that they regularly find text speak in submitted homework papers. Over 75% of parents said they had to edit and expand the abbreviated words found in their children's emails and text messages. Almost 89% said that this increase in the use of text speak is creating a language divide between their children and themselves.

The use of abbreviated, short-form language has even affected the way we speak. Apart from the everyday uses of "OMG", "FYI" and "BTW", I actually overheard someone the other day say "LOL" instead of laughing! So perhaps in future generations we will not only use the shortform versions of language to write, but they will become the way we express emotion.

# 2 WORK

## Are you on "self-distract"?

**One signature trait of our digital world is that it fragments our attention. Our attachment to our computers and hand-held digital devices has led us to view multitasking as our default setting, very often across several digital devices at once. Rather than view this as a negative, it makes us feel productive and efficient. In reality, though, multitasking actually makes us less productive and has been shown to diminish our ability to concentrate on one thing at a time.**

We are permanently distracted by a continuous deluge of emails, punctuated by social media notifications sliding into view, all competing for our attention and invariably interrupting our workflow. These incessant distractions compromise our productivity as we often end up making a small amount of progress over a multitude of tasks, leaving essential work unfinished.

According to new research by Clifford Nass, professor at Stanford University, who directs the CHIMe Laboratory (Communication between Humans and Interactive Media), which specializes in studying the interaction between human psychology and digital media, multitasking reduces our ability to distinguish between important and irrelevant information. Regular multitasking has been shown to actually reduce our capacity to complete tasks effectively because, as Professor Nass points out, we invariably end up focusing on the things we are not doing rather than on the task at hand.

He continues to explain that our inability to process information when multitasking has been shown to make managers less focused on their tasks and can lead to the making of ill-considered decisions.

In another study conducted by neuroscientist Professor Earl Miller, at the Massachusetts Institute of Technology, head scans were performed on a group of volunteers to monitor their brain activity while multitasking. The findings showed that when the volunteers were presented with a group of visual stimulants, the brain was only activated by one or two things at a time. Therefore, if the brain is overloaded with competing tasks, it has to alternate between them, causing its capacity to process information to be diminished.

This has been found to especially be the case when we try to perform tasks that are similar at the same time, such as answering a text while writing an email or talking on the phone, because similar tasks require us to use the same part of the brain. The result is that the brain becomes less able to function and consequently starts to function more slowly.

Even more surprising are the findings of Glenn Wilson, a psychiatrist at the University of London, that show that even just the thought of multitasking can cause the brain to slow down. He found that just by undertaking two simple actions at the same time, such as texting and emailing, your IQ can be reduced by 10 points! The knock-on effects of this can result in the type of mental fogginess usually caused by missing a whole night's sleep.

Multitasking has even permeated business culture in the form of corporate directives which actively encourage digital multitasking, setting internal time limits for generating email responses, green-lighting communicating via mobile phones on work-related topics within the work environment, and being permanently on call using internal "chat windows" for teams to use as another method of communication.

However, according to Professor Nass, the splitting of attention across a variety of screens and tasks impacts on the ability of employees to do their jobs effectively, with the greatest negative impact being on teamwork.

We need to re-learn how to "unitask", focusing fully on one thing at a time, immersing ourselves in the task at hand and not allowing ourselves to get distracted by the permanent alerts that accompany social media notifications and incoming emails. Completing tasks in a linear "one after the other" way as opposed to working on several things is the most productive and cohesive way to work.

For many of us who spend our days overwhelmed by the avalanche of emails in our never diminishing inbox, extending our working day into the evening seems like a sensible and productive solution. However, continuing to work from home late into the night with the aim of dealing with the tasks that we were unable to complete during the day because of regular distractions not only negatively impacts on our home life, but it has also been shown to dramatically reduce our ability to sleep well. This can lead to a state of exhaustion the next morning, followed by a dip in energy during the afternoon, which can further compromise our ability to work effectively.

According to a joint study on the effects of night-time digital activity outside of the workplace, conducted by a research team from the University of Washington, the University of Florida and the Michigan State University, the benefits of prolonged smartphone use for work purposes outside of the workplace and outside of working hours are negated by the reduced amount of time people then have to relax and unwind outside of the work environment.

The study also showed that work-related use of a digital device in the evening, and very often at night, was associated with interrupted sleep patterns resulting in fewer hours of sleep. The effects of this included a diminished capacity to use self-control in challenging situations, acute exhaustion and an overall feeling of being less focused throughout the day.

**"But we do have a choice. Our time is valuable and the way we spend it needs to be managed efficiently."**

There seems to be a direct correlation between how busy we are and how productive we perceive ourselves to be. Being busy makes us feel important and in demand. But we seem to be missing an important point. We can fill our time with numerous online activities that make us feel occupied, but we still end the working day feeling overwhelmed by our out-of-control, ever-filling inboxes and by the growing list of unfinished tasks that we still have to complete.

But we do have a choice. Our time is valuable and the way we spend it needs to be managed efficiently. In the same way as we plan how we spend our money,

# 6 steps to unitasking at work

**1 Allocate a time every day in your schedule to unitask**

In the same way that you schedule a meeting or conference call, block out a period of time for unitasking. Use this time to focus on special tasks that require your full attention. Select a specific period during the day when perhaps the workload is lighter or when there are likely to be fewer distractions.

**2 De-clutter your desktop**

While you are working, close all the screens that don't relate to the work that you are planning to undertake and open up one screen only so you can focus fully on the task in front of you. That will minimize the digital distractions competing for your time and allow you to work in a focused and methodical way.

**3 Turn off all digital alerts**

Disabling all digital notifications and alerts for specific periods of time substantially reduces the level of distraction experienced when working online and helps to facilitate your ability to focus on one task at a time.

### 4 Schedule a time to check your messages

Rather than checking your emails, voicemails and texts continuously as they appear, allocate yourself specific times during each day to deal with your messages. This reduces the number of interruptions to the other tasks that you are working on. Another effective technique for managing your email response rate is to add a note to your email signature indicating that email messages will only be checked periodically throughout the day, pre-empting the need for you to make an immediate response. Alternatively the same message can be sent via an auto-responder.

### 5 Tell your colleagues you are taking time to unitask

If you are in an open-plan space, a shared workplace environment or you operate an 'open door' policy in your own office, let your work colleagues know in advance that you need to work on an important task and require some uninterrupted time to complete the task at hand. They can be notified when you have completed the task and are available again.

### 6 Take an "unplugged" break

For those of us who are permanently distracted by the online activity of our desktops, laptops and smartphones, but who find the idea of disconnecting from our online world too challenging, there are apps that will do it for us. They can be set to turn everything off for a period of time which can be specified according to personal work schedules. Your digital down time can literally be used as a "breathing space" to take some deep breaths and reconnect with yourself.

separating it into different areas such as daily expenditure, monthly recurring bills and longer-term savings and investments, the time we spend online across various digital devices also needs to be managed by creating the equivalent of a personalized "income and expenditure" plan for our time.

There are ways to manage our online activity that will enable us to restore the balance between all of our digital worlds. We can implement self-generated systems to help us manage such activities by allocating a specific amount of time to each activity. By creating periods of uninterrupted time to fully engage with a task and other blocks of time to manage inboxes, messaging and phone calls, we can better control our impulse to respond to every email and alert as soon as we receive them. Even better, we can turn off on-screen alerts. We can allocate specific times for reading and responding to emails. We can set aside unplugged periods where we are distraction-free, to encourage creative flow and the generation of new ideas. We can turn our smartphones to silent mode during specific periods throughout the day and try to control our need to respond to everything immediately. We actually do not have to react instantly to anything. Just because we can doesn't mean we should. We have to take back control of our time. The smart option is to undertake an edit of your screen time so that you can spend it in a considered and effective way that allows you to prioritize what is really important.

Another consequence of our digital "over-connectedness" and our acute multitasking activities, and the lack of focus and attention that accompanies them, is that our people skills are also suffering thanks to a lack of practice. Taking our preferred route of using a managed, impersonal short form of communication such as texting or emailing rather than experiencing the emotional connection that getting up and talking to a colleague face-to-face provides can have a long-term negative impact on our emotional development.

## Work; don't work: the 90/10 rule

Extensive research into optimal working patterns has shown that to be our most efficient and effective we should change our level of activity every 90 minutes. Called the ultradian rhythm, researchers have found that 90 minutes is the optimal human limit for intensively concentrating on a single task. During a period of 90 minutes, especially when we're in full concentration mode, we move from a relatively high state of energy down into a physiological trough.

If the natural need for periodic rest in any extended performance situation is denied and ignored, this can lead to chronic stress and mind–body problems. The 10-minute rest period in between each 90-minute work period encourages the body to go into a "healing state" during which the disrupted ultradian cycle can reset. We seem to have become so adept at ignoring the obvious signals from our body that we need a rest, such as a lack of concentration, feeling anxious and irritable and general physical restlessness. We search ways to fulfil this need by turning to stimulants such as sugar and caffeine, causing our own stress hormones to go into overdrive and providing temporary bursts of artificial energy which are invariably followed by bouts of sluggishness and fatigue. By working instead with the body's natural cycles, we can increase our level of productivity while improving our overall healing and wellness.

**Yoga Snack**

### Eye-Strain Reliever

These two exercises together will relax your eyes and face, clearing your mind and restoring your focus.

**Eye Cup**

Rub your hands together to create some friction and heat. Cup your palms over your eye sockets while resting your elbows on your chest and bowing your head forward gently into your hands. This calms your mind and soothes screen-tired eyes.

**Lisa Sanfilippo**, yoga teacher

**Third Eye Point**

Right between the eyebrows you'll feel a bony notch. As you inhale, gently press this point with the pad of your index and middle fingers. As you exhale, release this point. Do this six times. It's a renowned point for alleviating anxiety used in both Indian acupressure and Traditional Chinese Acupuncture.

**Yoga Snack**

### Seated Side Stretch

You can do these at your desk or standing nearby. This brings a fuller, more enlivening breath into your body and oxygenates your blood more effectively, making it easier to focus if you're tired or lethargic.

1 Seated in your chair with feet on the ground, press your hips down strongly and hold your left hand to the edge of the chair.

2 Lift the right arm up, and reach over your head towards the left side, stretching the side flank of your body.

3 Breathe deeply and evenly from the bottom of your ribcage and use your breath to expand your lungs and stretch your intercostal (between the ribs) muscles.

4 Do this for 5–10 rounds of breath on the first side, and then switch to the other side for the same number of breaths.

**Lisa Sanfilippo**, yoga teacher

## Step away from the desk

The most obvious way to take a 10-minute break between your 90-minute work cycles is to leave your desk! Actually get up and walk around. Talk to your co-workers – surprise them by making a personal appearance! Engage in some real face-to-face conversation or find a quiet place to practise some mediation and deep-breathing techniques or even take a moment to have a "Yoga Snack" – doing one of Lisa Sanfillipo's bite-size yoga exercises.

A 10-minute break away from your desk gives your mind a chance to "power down", so that it can recharge, and your body an opportunity to get more active or wind down to a relaxed state, so that you can reconnect with yourself and return with a clear mind and a fresh perspective.

A survey undertaken by Salary.com contained several shocking statistics: 89% of people taking part in the survey admitted to wasting time every day while at work, an increase of over 20% on the previous year's results. The rise in the percentage was found to be due to an increase in the amount of time wasted on non-work matters.

The survey provided the following analysis of the time wasted on non-work activity every day:

- 31% waste on average 30 minutes per day
- 31% waste approximately one hour every day
- 16% waste approximately two hours every day
- 6% waste on average three hours per day
- 2% waste approximately four hours every day
- 2% waste five hours or more per day

The survey concluded that 4% of the sample group surveyed wasted at least 50% of their working day undertaking non work-related activities, with 26% of the group admitting that their greatest timewasting activity was browsing the Internet during working hours.

Apart from the obvious consequences of working time lost and the negative impact on productivity and work performance, those hours spent at a desk surfing the Internet on non-work activities are seriously detrimental to our health, as they prolong the time we spend sitting down and inactive.

**Yoga Snack**

### Full Body Wall Stretch

This is similar to yoga's Downward Facing Dog pose but a bit easier to manage in an office. It stretches the shoulders and the long muscles that line your spine, and improves circulation in the lungs, clears your head and stretches tight hamstrings.

1 Press your hands into the wall just below shoulder height. Straighten your arms and bow forward.

2 Walk your feet back just beneath and behind your hips. This makes an L shape at the wall. Direct your breath towards and into your lower back for 5 breaths.

**Lisa Sanfilippo**, yoga teacher

## Yoga Snack

### Recharge Pose

This pose is simple and incredibly powerful. Ten minutes with your eyes closed, relaxing, will help to press the "reset" button on your energy levels.

1 Find an unused room where you'll have 10 minutes' privacy. Find a stable chair and, facing the seat, lie down on the floor on your back. Place your calves and ankles on the seat of the chair, with your hips and lower back relaxing on the floor.

2 As you breathe gently allow your neck, shoulders, upper and lower back to relax towards the floor.

**Lisa Sanfilippo**, yoga teacher

### Sitting is the new smoking

According to research by the UK government organisation "Get Britain Standing", the average time spent sitting at work every day in the UK is 8.9 hours. Their research shows that "after just 90 minutes of sitting our metabolism shuts down."

Prolonged sitting in the same position stagnates the body thanks to the lack of movement. Muscles that should be contracting are static and without muscle contractions, which help to remove toxins from our body, stimulating both the lymphatic system and blood flow throughout the body, fats such as triglycerides and sugars in the form of glucose remain in our system instead of being filtered.

According to several studies reviewed by *The American Journal of Preventative Medicine* there is further compelling evidence that sitting for longer than four hours every day can also lead to obesity, as the enzymes responsible for burning fat within the body shut down, reducing the body's metabolic rate and also adversely affecting the blood flow in the legs. These extended periods of inactivity can also cause an extensive list of other potentially life-threatening diseases including:

- Heart disease
- Diabetes
- Obesity
- Muscular and back issues
- Deep-vein thrombosis
- Brittle bones
- Depression
- Dementia
- Cancer

Prolonged sitting and physical inactivity has been ranked in fourth position on the World Health Organisation's list of life-threatening activities, overtaking obesity. The recommendations from the World Health Organisation are that adults need to undertake regular exercise, ideally 30 minutes a day, five days a week – or a minimum of two and a half hours of moderate exercise spread over a seven-day period. However, the Organisation warns that although incorporating a regular exercise routine into your schedule will help to limit the effects of time spent sitting, it won't protect you from the more serious and damaging health issues that are associated with the general lack of activity that accompanies a sedentary lifestyle.

According to Dr Michael Jensen, of the Mayo Clinic, there are some practical actions we can take to counteract the long-term effects of prolonged sitting.

- Try to stand at work at least part of the time during the working day. There are now "standing desks" that enable you to work while standing. If your workspace does not have standing desks available then make sure you get up and move around every 30 minutes if possible, for a few minutes. Movement is crucial to our physical wellbeing – just standing without actively walking around is not going to get the blood circulating through the body enough to combat the negative effects of prolonged sitting. Instead of using the office chair supplied by your workplace try using a stability ball, as they have been shown to be beneficial in engaging the core muscle groups

- Use every opportunity to try to take the stairs rather than using a lift (elevator) and walk over to talk to your co-workers rather than sending them a message while in a shared workspace. Keep your smartphones and other digital hand-held devices near you but not within reach so that you have to get up on a regular basis to use them. When making or receiving a call on your hand-held digital device use the opportunity to stand up and walk around. Invest in a digital pedometer that will monitor the amount of steps you make every day. The average number of steps a sedentary person takes can be as little as 1,000 to 3,000 a day. Set yourself a goal of increasing your steps every day by 500 each week until you reach the 10,000 steps per day recommended by the Surgeon General of the United States.

## Out to lunch

One of the prime opportunities for us to step away from our desks, leave our screens and get moving is during the lunch break. However nearly 70% of us eat lunch at our desks, which may make us feel as if we are being productive by minimizing the time spent not working, but in reality we can gain so much more by taking a restorative, unplugged break from our desks. Instead of posting updates on social media, doing some online shopping or checking out the latest viral YouTube video while eating a sandwich, we can use this time to unplug for an hour, get up, leave our desks and go somewhere.

A change of environment, combined with physical activity – whether it is walking to another area of the workplace or actually leaving the building to go outside for a short walk or workout – will restore energy levels, sharpen focus and clear the mind. That short time away from your work environment will leave you more energized, focused and relaxed, ready to bring your best self to the conference table.

Every day could bring a new lunchtime experience. Experiment with different activities, explore new places in the area, try new foods and use this time as a gift to yourself. A time for you to reconnect with yourself.

If we needed a reminder that having a screen as a lunch companion is actually really sad, a recently launched product really drives the message home. The latest invention designed to service our incapacity to unplug even during meal times is the "anti-loneliness ramen bowl", a noodle or soup bowl that has been designed with a slot for your smartphone so that you can eat while looking at your screen. When not in use as a bowl, the unit doubles up as a smartphone dock with the bowl serving as a sound amplifier. Is this the ultimate antidote to mindful eating, encouraging us to eat mechanically while watching a screen? Or is it how we are all going to be eating in the future?

## Digital rules of engagement

### The new business protocol

We all accept that the smartphone is now at the epicentre of business communication. It acts as our centralized information hub, providing immediate access to our co-workers and clients by enabling us to email, send text messages, research and compile reports, and make calls. It is clearly an indispensible tool for connecting a company's workforce and there is no denying that it increases levels of efficiency, productivity and communication.

However, smartphone use without any corporate boundaries or recommended guidelines can have a negative impact on corporate culture, focus and productivity. Nowhere has this become more evident than during meetings. Although both smartphones and tablets have become fundamental to connectivity in the workplace, the jury is still out as to whether they have a place in meetings.

### Meet and tweet

Smartphones and tablets have replaced the notepad and pen in meetings and they appear on conference tables throughout our corporate culture as a familiar accompaniment to the ever-present bottles of mineral water.

Some companies, however, have adopted a blanket ban on the use of digital devices during meetings. But in order to avoid "throwing the baby out with the bath water", in our attempts to establish an implementable digital business protocol, we need to differentiate the use of smartphones from the use of tablets and laptops during meetings, as there is a difference in the perceived use of different types of digital devices in a meeting context.

Unlike laptops and tablets, smartphone are usually associated with texting, tweeting, checking social media and other non-work activities. Bringing a smartphone into a meeting can signal to your colleagues that your focus is elsewhere and you are not giving your full attention to the matters at hand, particularly if you periodically slide your smartphone under the conference room table to check your messages while someone is speaking. Using a smartphone during a meeting, however surreptitiously, is generally viewed as showing a fundamental lack of respect for the meeting itself and the other people present.

If the taking of your smartphone into a meeting cannot be avoided, owner of Princeton Public Speaking, Matt Eventoff, recommends that as soon as you take the device out in a meeting, you should inform everyone present that you use your smartphone or tablet as a note-taking device.

Generally, tablets and laptops are viewed as more appropriate digital devices for note-taking in meetings than the smartphone. However, using a laptop can mean that the upright screen acts as a visual barrier to the other people attending the meeting, so overall tablets seem to be viewed as the most appropriate and least offensive digital devices to use in a meeting context.

The University of Southern California's Marshall School of Business undertook a survey which formed part of its Global Mobile Research Program, an annual study of mobile (cell) phone users and how they use and value their mobile phones. Their survey found that of the study group, made up of 5,500 North Americans, 84% thought texting or writing emails during a formal meeting was inappropriate behaviour; 75% thought the reading and checking of incoming messages and emails was not acceptable behaviour in a meeting context; 66% thought it was also inappropriate to be texting or emailing even at informal meetings; 86% found the taking of phone calls during a meeting completely unacceptable, while 22% thought that using a smartphone during any form of meeting was completely inappropriate.

Roger Lipson, founder of The Lipson Group, found during the course of his survey of executives' smartphone and tablet use in meetings that the opinions of his survey group as to whether using digital devices in meetings was acceptable behaviour varied immensely according to the age group of the respondents.

As expected, Millennials are 300% more likely to think that checking text messages and emails during informal meetings was acceptable professional behaviour, compared to respondents over the age of 40. It was also noted that the disparity between the opinions of Millennials and those of the older respondents, who are invariably their seniors and have influence over their careers, could in the long term negatively impact younger people's career advancement.

Perhaps in the same way that "dress down Fridays" have been incorporated into the working week to encourage a more relaxed interaction in the workplace, a weekly "unplugged day" could also be introduced into the working week. This would act as a catalyst for creative thinking, face-to-face meetings and group interaction encouraging unitasking and creative flow.

Clearly there has to be flexibility as to how each company integrates their particular unplugging policy into their corporate infrastructure in order to derive the maximum benefit for company employees. A customized approach that addresses the global digital overload issues within the company's workforce, and the related stress level that this places its employees under as they try to cope with the overwhelming deluge of emails and messages, can only benefit the productivity of the company and the general wellbeing of its workforce.

Companies are beginning to establish some guidelines and rules for digital connectivity in order to manage digital workloads within the workplace. Employees are being encouraged to disconnect and stop feeling as if they have to be on call 24/7, and to learn to place a greater value on their "digitally dark" time. Companies are realizing the benefits of allowing employees to balance their work-life schedule and are encouraging them to review their personal work-life divide. Employees' "digital downtime" will transform into a positive opportunity for them to reconnect with themselves and encourage them to learn new coping techniques for controlling digital overload, which in the long term can dramatically increase productivity, reduce working days lost through stress-related illness and demonstrate a corporate culture that values its workforce.

Industry is now beginning to realize the benefit of corporate unplugging, and companies are starting to implement their own unplugging schedules. German car manufacturer Volkswagen is scheduling email shutdowns for some employees 30 minutes after their work day ends. BMW is planning to implement a policy that prevents their workforce from being contacted after hours and Goldman Sachs is encouraging junior employees to take time out and have weekends off.

Quirky, a New York-based start-up that brings new inventions to market, has instituted a policy of a "blackout week" once every three months, during which time no one other than the employees handling customer service enquiries are permitted to work, to avoid the possibility of employees checking emails. Quirky CEO explained the reason behind the quarterly blackout was that people were getting exhausted and burned out sitting at their desks all day, and that they needed to see and experience other things.

Arianna Huffington, president of media giant the *Huffington Post*, has implemented a new internal email policy to encourage employees not to email after hours and have some down time. She has also set up two spaces in their New York offices for employees to be able to nap during the working day.

Companies are beginning to recognize that people now have so much more digital data to process that they are suffering from information overload. This, according to a 2010 study by New York research company Basex, has led to corporate America losing close to $1 trillion in lost work hours through employees using time at work to answer non-work emails and messages during the working day. Employees are present but are not being productive.

The motivation behind these new corporate mandates clearly has as much to do with employee welfare as with the recognition that encouraging employees to have some down time away from work actually increases their productivity and therefore the corporate bottom line.

## Are you getting a reputation?

We all have a digital footprint and therefore a digital reputation. In our over-enthusiastic need to share every aspect of our lives by posting "in the moment" experiences via a range of social media, it is often hard to remember that those moments of fun can have a long-term damaging effect on our professional digital reputation, which is permanent.

While it is widely accepted that prospective business contacts and employers will check out our online presence before making any form of direct contact with us, our attempts to edit or erase our embarrassing moments cannot be guaranteed to work. Our desire to retrospectively "cyber-wash" our online reputation can only be effective to a degree, as we cannot control how many times the information that we posted has been shared, copied and viewed by people unknown to us.

We offer up personal and private moments

voluntarily and willingly to a digital infinity that has no tracking device. Our digital footprint is there for anyone to access, and this can be both a positive if our online presence is managed well, or a huge negative if it is mismanaged, as ultimately our online reputation forms an intrinsic part of our path to professional success.

## Guidelines for the responsible use of smartphones in meetings

- Many companies have implemented a blanket "no smartphone" rule for all meetings. One company found a witty, lateral solution to dealing with smartphone use in meetings, clearly inspired by a scene in a spaghetti western. They placed a wicker basket at the entrance to their main conference suite, accompanied by a sign which featured a picture of a smartphone with the message "Leave your guns at the door."
- For companies who find a no smartphone rule too stringent a measure to implement successfully, or encounter too much resistance to it from their workforce, a less extreme option is to establish some ground rules prior to the start of the meeting by asking everyone present to turn off their smartphones or tablets (unless used for note taking) so that they can give their full attention to the matters being discussed.
- Another option that has worked its way into corporate culture is the incorporation of "texting breaks" during meetings, to allow participants to either step out of the room at a given interval period during the meeting or to remain in the meeting room and switch on their digital devices for a short period of time, alleviating the need for clandestine smartphone checking under the desk.
- Sometimes meetings coincide with other urgent matters, so if you think you are going to have to take an urgent call, the correct procedure is to let the other people in the meeting know before the meeting starts that you are expecting a call and set the phone to vibrate, disabling the ringer for minimum interruption. Once you receive the call, excuse yourself and leave the room to take the call.

## Case Study

### Arianna Huffington,
*editor-in-chief of Huffington Post Media Group*

Arianna Huffington is president and editor-in-chief of the Huffington Post Media Group. As the co-founder and editor-in-chief of the Pulitzer Prize-winning *Huffington Post*, Huffington has also written 14 books, and is a nationally syndicated columnist and radio host. In recent years she has shifted her focus to her interest in personal and spiritual wellbeing. Her latest book *Thrive: The Third Metric to Redefining Success and Creating a Life of Well-Being, Wisdom, and Wonder* instantly became a bestseller, topping the *New York Times* bestseller list. In 2013 Huffington, aware of our growing addiction to our digital devices, set a challenge for her readers in the *Huffington Post* to try to spend at least 30 minutes per day, every day, unplugged.

**1 Do you think that our need to be connected digitally 24/7 is actually a choice that we make?**

Sometimes it's a choice, but more important, it's part of our collective delusion that being always connected is the necessary price for achieving success. And as Kelly McGonigal, a psychologist who studies the science of self-control at Stanford's School of Medicine, puts it: "People have a pathological relationship with their devices. People feel not just addicted, but trapped."

**2 Do you think our addiction to our digital devices is symptomatic of a deeper malaise within our society born of our inability to just "be" with ourselves?**

Yes, our hyperconnectedness is the snake lurking in our digital Garden of Eden. We are finding it harder and harder to unplug, renew ourselves, and make a real connection with ourselves and others. The first stages of the Internet were about data and more data. But now we have plenty of data – indeed, we're drowning in it – and all the distraction we could ever hope for. Technology has been very good at giving us what we want, but not always what we need.

**3 In a professional context what do you think are the greatest negatives of our permanent digital distraction?**

To give just one example, our relationship with email has become increasingly one-sided. We try to empty our in-boxes, bailing like people in a leaky lifeboat, but more and more of it keeps pouring in. How we deal with our email has become a big part of our techno-stress. It's not just the never-ending e-deluge of emails we never get to – the growing pile that just sits there, judging us all day – but even the ones we do get to, the replied-to emails we think should be making us feel good. And the problem is that with smartphones, email is no longer confined to the office. It comes with us – to the gym, to dinner, to bed.

**4 What did you find to be the positives from being "unplugged" during the digital detox you undertook in 2013?**

Unplugging meant rediscovering and savouring the moment for its own sake. Which is to say, taking in a view without tweeting it. Eating a meal without Instagramming it. Hearing my daughters say something hilarious and very shareable without sharing it.

**5 What would be your three best recommendations for achieving a successful live/work balance in our digital world?**

I have 12 steps I recommend in *Thrive*, and each one of us needs to pick the step that most resonates with us. Here are three:

- Have a specific time at night when you regularly turn off your devices – and gently escort them out of your bedroom. Disconnecting from the digital world will help you reconnect to your wisdom, intuition and creativity. And when you wake up in the morning, don't start your day by looking at your smartphone. Take one minute – trust me, you do have one minute – to breathe deeply, or be grateful, or set your intention for the day.
- Introduce five minutes of meditation into your day. Eventually, you can build up to 15 or 20 minutes a day (or more), but even just a few minutes will open the door to creating a new habit, so you can enjoy all the many proven benefits it brings.
- Drop something that no longer serves you. I did a major "life audit" when I turned 40, and I realized how many projects I had committed to in my head – such as learning German and becoming a good skier and learning to cook. Most remained unfinished, and many were not even started. Yet these countless incomplete projects drained my energy and diffused my attention. As soon as the file was opened, each one took a little bit of me away. It was very liberating to realize that I could "complete" a project by simply dropping it – by eliminating it from my to-do list. Why carry around this unnecessary baggage? That's how I completed learning German and becoming a good skier and learning to cook and a host of other projects that now no longer have a claim on my attention.

**6 I read with interest in your new book *Thrive* that you view "sleep" as one of the most fundamental yet underrated "stress-busting and life-enhancing activities". With our digital over-connection, going to sleep with a smartphone next to the bed, guarantees a disturbed night's sleep. Do you have a personal rule for smartphone use in the bedroom apart from not charging it there?**

Yes. My bedroom is a device-free zone. I read real books in bed and, of course, I charge my phones far, far away to avoid the middle-of-the-night temptation to check the latest news or emails.

**7 Running a hugely successful media empire while at the same time preserving your human side by prioritizing personal interaction, such as sending a personal note, sharing the things that make you vulnerable, is a complete counterpoint to the professionally perfect, ironed, edited "public" version of themselves, usually associated with most successful entrepreneurs. Yet you chose to share your human side. Do you view that as a measure of personal success as a human being or actually an intrinsic part of professional success today?**

Yes, there is a direct connection between our personal and professional success. That's why we need to do everything we can to protect and nurture our human capital. My mother was an expert at that. I still remember, when I was 12 years old, a very successful Greek businessman coming over to our home for dinner. He looked rundown and exhausted. But when we sat down to dinner, he told us how well things were going for him. He was thrilled about a contract he had just won to build a new museum. My mother was not impressed. "I don't care how well your business is doing," she told him bluntly, "you're not taking care of you. Your business might have a great bottom line, but you are your most important capital. There are only so many withdrawals you can make from your health bank account, but you just keep on withdrawing. You could go bankrupt if you don't make some deposits soon." And indeed, not long after that, the man had to be rushed to the hospital for an emergency angioplasty.

**8 Do you think the Internet has also served as a positive counterpoint that has changed and humanized us, through its levels of accountability and transparency?**

Yes, and not only through accountability and transparency. Paradoxically, one of the biggest growth sectors for tools to help us deal with technology and hyper-connectivity is . . . technology. Many in the tech world have realized there's a growth opportunity for applications and tools that help us focus and filter all that data and distraction. At the end of *Thrive*, I have compiled some of my favourite tools that can help us. As Steve Jobs said, "Focusing is about saying no."

## Case Study

### Lewis Lapham,
*founder of Lapham's Quarterly*

Lewis Lapham was the editor of *Harper's Magazine* from 1974 to 2006 and acted as its managing editor from 1971 to 1975, after having worked for the *New York Herald Tribune* and the *San Francisco Examiner*. He is mainly credited for the contemporary design and prominence of the magazine and became editor-at-large in spring 2006. Lapham left *Harper's* in 2006 to found the *Lapham's Quarterly*, a literary magazine that he first conceived in 1998. In 2007 he was honoured by his peers by being inducted into the American Society of Magazine Editors' Hall of Fame.

Lewis Lapham is unplugged. He doesn't have a computer, smartphone, use email or surf the Internet. He's not on Facebook and doesn't own an iPad. He does admit to sparingly asking the younger members of his editorial team to "look something up online". He accepts that the digital world has an extraordinary side to it, filled with possibilities, but it is the downsides that concern him most.

He believes that we are voluntarily handing over our own thinking and creativity to a computer; "a machine that doesn't speak". He grew up with words and books and still takes pleasure in writing with a pen on paper, describing it as a "sensual" experience. For him, writing on a digital device is not writing for a human being; he views it as "writing for an algorithm". He also believes that the content we create online has no intrinsic value as its value is now determined by the volume of people that read it. Lapham goes on to share a quote by Max Frisch: "Technology is the knack of so arranging the world that we don't have to experience it." He shares this view that through our digital overload we are living in a virtual reality bubble within which we are losing touch with each other, and ourselves.

For Lapham language is everything; he cannot understand "how we can discover a life of our own without the language to tell our story". This directly refers back to the philosophical question: "If we don't have a word for it, can we think it?" Lapham has a mission to undertake his own literary version of an archaeological dig to uncover the wisdom to be found deeply buried beyond words. He stands by his belief that language is infinitely more powerful than the Internet, and wants to demonstrate that literature has the power to transform, inspire and enlighten. For Lapham it's not so much about the medium but the message and for him digital communication definitely gets lost in translation.

I had the pleasure to interview Lewis Lapham in person at the offices of *Lapham's Quarterly* in Gramercy, New York. His glass-walled office housed a temple to literature shared only by his desk and a library-worthy array of books. The books on his current reading list were stacked on his desk, because the desk wasn't the empty sterile platform usually reserved for a computer or a laptop. Not having one of these, Lapham spends his time reading, writing in longhand and thinking. As I started to ask him the questions I had prepared for him, he handed me a copy of the *Lapham's* spring 2012 issue entitled "Means of Communication". The issue is a kaleidoscopic, textured, complex and multi-layered commentary on how we communicate, with references spanning Charles V to Lenny Bruce, via Toni Morrison, Schopenhauer and Helen Keller.

# Q&A

**1 What do you view as the greatest downside to our digital over-connection?**

It destroys the value of time. We lose the ability to experience the richness and quality of our own lives. Without time to know ourselves we end up with a glance at everything and an understanding of nothing.

**2 Do you think the medium impacts on the message?**

I asked Lapham about his method of writing "unplugged" without a computer. He explained that it was somewhat of a process. He first writes what he wants to say in longhand with a pen on paper, then dictates what he has written into a small voice recorder from which his PA transcribes a manuscript for an email.

**3 Do you believe that by using reductive language we reduce our capacity for thought?**

What we lose is language. What survives over time is the force of the human imagination and the power of its expression. The Internet doesn't trade in those commodities; it traffics in data, in a medium reduced to the vocabulary of a child's alphabet blocks. Lapham cited the premise of Marshall McLuhan's *Understanding Media*, published in 1964, "we become what we behold...we shape our tools and thereafter our tools shape us." Content, Lapham said, follows form; new means of communication give rise to new structures of feeling and thought, which tend to be accelerating backwards in time towards a primitive shaking of rattles and beating of drums. He quotes Toni Morrison's 1993 Nobel Prize speech in Stockholm, "Word Work is sublime because it is generative, its felicity in its reach toward the ineffable...we die, which may be the meaning of life, but we do language, and that may be the measure of our lives."

**4 How do you think the virtual world we have created impacts on our real world experience?**
Americans like shiny new things that move. We don't like silence, stillness, being alone with our own thoughts. We search out distractions. Lapham feels that we check ourselves into the nearest online cage because we are terrified of freedom, feel safe only among the buzzing of bees in a hive.

**5 What are the benefits of being unplugged for you?**
Being unplugged gives me the freedom to explore and discover without having to follow instructions, without being told to fill out the forms. For me the greatest benefit of being unplugged is having time for yourself, your own time instead of marching to the tune of other people's time, and opening up and out instead of a pinning things down.

**6 Much of the content posted online now is visual. Do you think we are becoming too focused on these commercial advertisements for reality?**
The visual image addresses a passive spectator. The imagination is not engaged in the same way that it would be with the word, whether written or spoken. It is the labour of the writer turned to the wheel of the reader's imagination that gives rise to the power of literature. Television induces a state of drugged absorption, which is lifeless.

**7 Facebook presence is often an edited, digitally enhanced perfected version of an imaginary self, but isn't it our flaws that make us human?**
Our flaws are what make us human. When we cut and paste digitally enhanced artificial replicas of ourselves, we are trading in a counterfeit currency.

**8 If you could make one change to the way we communicate now, what would it be?**
Unplug one day a week, take our attention out for a walk, read a book, cross a bridge, fondle a loved one, climb a tree, swim in a lake, maybe learn that there are more things in heaven and earth than are dreamed of by Walt Disney or Time Warner Cable.

# 3 PLAY

## Relationships

### I "*likes*" you

**We are becoming increasingly aware that being digitally connected 24/7 impacts on every aspect of our personal relationships. Our connection to our smartphones is replacing our ability to connect with each other. We are now spending increasing amounts of time "together alone", surrounded by friends and family but looking at our smartphones instead of at each other. We also seem to have become an instant "digital response squad", checking and responding to emails, texts and posts on social media as soon as they hit our digital devices. Including with increasing frequency during the night and then first thing in the morning – and even, shockingly, during sex.**

A commentary on the oppressive nature of our smartphone addiction, its domination over every aspect of our lives and its negative and destructive effect, particularly on relationships, was recently highlighted in a YouTube video written and created by Charlene de Guzman. The catalyst for creating this video came about while she was at a concert and everyone around her seemed to be recording the show with their phones instead of actually watching the band. De Guzman explains, "That's when I started to realize how ridiculous we are all being, myself included."

De Guzman's video, called simply "I forgot my phone" makes for uncomfortable viewing. It's a hard-hitting commentary on our obsession with our smartphones. The video follows de Guzman for a day, covering a series of events shared with her partner and her friends, where she is the only one without a smartphone, throwing into sharp focus how our addiction to a 2-inch screen negatively impacts on every aspect of our daily lives. The clear message of this personal video lament is that we seem to be spending more time *viewing* life than *living* it. Her message clearly touched a raw nerve for the millions who saw it – to date it has had almost 50 million YouTube views. (http://www.youtube.com/watch?v=OINa46HeWg8)

**"We are now spending increasing amounts of time together alone."**

The isolation that our connectivity, particularly on social media, can create was explored in a YouTube video intentionally posted on Facebook by Mauricien, a community site for Mauritius. It is a powerful and poignant take on our replacement of love with "likes" and scenes with screens, and starts with the statement, "I have 422 friends and yet I am lonely. I speak to all of them every day, but none of them really know me." The idea behind this video piece was to explore, interestingly by using contemporary prose, the emptiness of our digital over-connection. The visual narrative explores all the real moments of life that we are all missing by not looking up and never stepping away from our smartphones. Those moments are described as the "spaces in between" looking into someone's eyes or at a name on a screen. The video's aim was for it to be watched at least once by everyone on Facebook; it seems to have genuinely struck a chord with Facebook users, as to date it has had over 60 million views.

## Second screening

Our inability to concentrate on one thing thanks to the demands of technology has been established. We seem to be unable to have a conversation without checking our smartphones, or answer an email without tweeting at the same time – and it now seems we are no longer able to just watch TV. Before the introduction of the life-altering smartphone, the term "multi-screen" used to be most readily associated with a large cinema complex. Now multi-screening is our personal default setting for how we use our digital technology. Whether at our laptops or desktops, or using our mobile digital devices, we permanently have multiple screens open, switching from one to another continuously. Just doing one activity at a time in a mindful and focused way seems to feel slow and unproductive. The same seems to apply to watching TV. According to a study by Pew Research, as a part of their "Internet and American life" project, over 50% of mobile phone users in the US use their digital devices while watching television. This practice has been called "Second Screening" and is so widespread that several TV networks both in the UK and the US are actually using it to not only provide supplementary information to support programming but also to encourage viewer interaction via their network websites and social media platforms.

According to the findings of the Pew survey, a study that sampled 2,250 American young adults aged between 18–24 years, mobile and smartphone users used their devices to contact friends on social media and look up information in real time, mainly during the TV advertisement breaks.

About a quarter of the study group sent text messages to friends watching the same TV show elsewhere, while about 20% used their mobile phones to look up content mentioned on the programme they were watching. Around 11% posted comments online as they were watching the show, with 38% engaging in the most popular use of mobile phones while watching TV – playing games on their phones during advertisement breaks.

## Size matters

I think part of the smartphone addiction problem is its size. We can hold our devices in one hand. They are a slim and sleek portal to our digital life; superficially they look discreet but in fact they command our full attention. The practicalities of our daily lives are facilitated by our ability to organize them seamlessly, using a plethora of apps on our smartphones. We carry all our information with us: our contacts, our banking, our emails, our projects, our photos and our social media platforms in a single portable hand-held digital device.

To carry the same information in the pre-smartphone era would have required us to carry our computers or laptops and bulky mobile phones with us everywhere. With our smartphones we can now organize our life. We can communicate with and check our homes by remote. We can restock our fridges and make dinner reservations via Open Table. Once we've arrived we can check in on Facebook or Foursquare. We more often than not take a picture of our meal, so that we can post it on Instagram or Twitter. After the meal we can immediately post a restaurant review on Yelp and finally book a taxi to take us home using the Uber app. We can continuously monitor our health, our work, our finances and our social popularity. All of which leaves little time for real, face-to-face offline relationships. It brings to mind the statement, "Just because we can doesn't mean we should."

According to mobile technology authority Toml Ahonen, checking our smartphones continuously can actually ruin our relationships. Whether in a one-to-one situation or in a group context, the sight of illuminated faces interacting with their smartphones, ignoring the people around them, has become the default setting for any meeting or social gathering. We are not fully present, with our time and attention being permanently hacked into by the demands of managing our digital life rather than our real life. We are losing the ability to pay attention to each other, to completely focus on one another, without the constant interruption caused by checking our smartphones. The sight of a couple at dinner in a restaurant, each of them with their heads down texting, has become accepted as the norm. Let's hope that at least they are texting each other!

Recently I overheard a conversation between two Millennials bemoaning the fact that they had gone out clubbing the previous night and were so bored because no one was interacting with anyone, as everyone was standing around the dance floor, heads down, looking at their smartphones.

If smartphones were the size of a vintage dial telephone, a comparatively giant box with a phone receiver and round dial, would we still think it was OK to place one – or even worse, two of them – on the dinner table between us and our guests? If it wasn't for their reduced scale and minimal design, would we not think that putting a telephone on the table in a restaurant was at best bizarre behaviour and at worst totally disrespectful? Even though the discreet design of our smartphones makes them appear less intrusive, the reality of their presence at the dinner table does not in any way diminish the demand they make on our time or our attention. The fact that they are palm-sized seems to make us think that checking them every few minutes is somehow acceptable.

The demands of our digital devices on our time and for our full attention have now taken priority over giving our attention to the people we are with. We all know how lonely it feels to be with people who are with us physically in the same space but who are not really present. Either they are not paying us any attention at all, as our window of opportunity to interact has been hijacked by a smartphone, or more often than not we end up at the receiving end of "skimming" – a form of soundbite surfing where the person you are with picks up every third word because they are not really paying attention, as they are just going through the motions of listening but are really concentrating on texting or scanning through emails.

According to Rapid Change therapist Howard Cooper, "People have a deep psychological need for significance, we need to matter!"

He believes, "Being connected with the world or imprinting our views for all to see [on Twitter, Facebook etc] is a way of gaining significance, of having a voice that is heard. If we get together to have a chat and 'really connect' with someone and then they proceed to spend most of their time on their phone, we in turn, mostly as a self-protective mechanism, look to fulfil that need for significance elsewhere, and generally it is our phones that we reach for. They are our first port of call when we need to redress the balance and want to feel significant or at the very least digitally connected to someone, when our real live face-to-face connections fail. This can then become a self-perpetuating cycle when the other person momentarily looks up from their screen and sees you NOT connecting with them and so they return to their digital device to fulfil their own need for significance." So this destructive, self-perpetuating, non-communicative cycle continues.

## Read alert

The act of using a smartphone while in the company of someone in a social setting has reached such epidemic proportions that it now has its own term, "phubbing". An amalgam of the words "phone" and "snubbing", the term is used to describe someone who checks their phone while they're with others, ignoring them in favour of their smartphone.

The dramatic increase in phubbing unsurprisingly has a direct correlation with the increase in smartphone usage and our use of social media. We seem to find it difficult to be in any social situation without either taking a series of selfies to post on Facebook, snapping the food we are eating and uploading it to Instagram, or simply texting and answering emails while in the company of others.

Imagine if the alerts we get on our smartphones every time we receive a message took the form of an actual "snail mail" letter that arrived at the table while we were with other people. Following that thought through, what if – given the fact that in the space of an hour we could receive a substantial quantity of emails – that meant we would receive a whole pile of letters that were placed on our dinner plates? I imagine the normal response would be to gather them up and put them away for reading later, rather than proceeding to open each letter one by one and read it in front of everyone else at the dinner table, which would be considered at worst to be extremely rude and at best really bad manners.

But that is exactly what we are doing when we check our emails, send texts and post on social media while we are in the company of others. When did we forget how to behave socially in a civilized and respectful way?

**"Imagine if the alerts we get on our smartphones every time we receive a message took the form of an actual 'snail mail' letter that arrived at the table while we were with other people."**

Alex Haigh, a 23-year-old Australian, was so appalled by this new epidemic of texting while in company that he launched the "StopPhubbing" campaign in 2013. It has clearly captured the zeitgeist as it has rapidly become a global movement. (www.stopPhubbing.com)

According to the statistics he has posted on his website, on average a restaurant will witness 36 cases of phubbing during a single dinner sitting; 87% of teenagers admit that they would rather text than have to talk face-to-face, and if phubbing were a viral epidemic it would devastate the population of China six times over. The StopPhubbing campaign has a clear message: "Leave your phone in your pocket and have a chat with the real world." They have even produced some downloadable posters with their witty anti-phubbing messages – my personal favourite being one especially designed to help eliminate phubbing from restaurants. "While you finish updating your status, we'll gladly serve the polite person behind you."

Perhaps it's time for phubbing to be banned in the same way that smoking in bars and restaurants was banned. After a little resistance we now accept "smoke free" zones as the norm, so perhaps it is time for a new "mobile free zone" policy to be implemented in restaurants, bars and clubs so that face-to-face interaction becomes the accepted mode of communication once more.

Our general lack of digital manners and the anti-social behaviour caused by our unchecked overuse of smartphones also has another unexpected, yet equally damaging, consequence: it causes a significant increase in our carbon footprint. Given its slimline proportions, the smartphone consumes far more energy than we realize. Those of us who use a wireless connection to access a minimum of 1.5 GB of data each month incredibly use more energy each year than an average-sized fridge. So if the anti-social aspect of phubbing doesn't get us to re-evaluate our digital manners, perhaps our impact on the environment will!

**"The act of using a smartphone while in the company of someone in a social setting has reached such epidemic proportions that it now has its own term, 'phubbing'."**

## Unplug and play

The fact that we seem to be losing our social skills by continuously prioritizing our smartphones over the people we are with inspired Polar, a Brazilian beer brand, to come up with an ingenious solution: a beer cooler that cuts out all Wi-Fi, 3G and 4G signals and GSM within a five-foot radius.

Placed strategically in bars, restaurants and clubs, a large bottle of beer sits in the cooler in the middle of each table and instantly disables all digital communication around it. Polar believes, radically, that we should actually enjoy each other's company instead of having our heads down looking at our mobile phones! As we've already discussed, restaurants, clubs and bars globally are filled with people standing or sitting together all staring at their digital devices instead of being present. It's a problem that is widespread and clearly no respecter of location or environment, as to some degree it affects us all.

To launch the beer cooler concept, Polar teamed up with Brazilian agency Paim

Comunicação to create an ironic video that shows, using a case study, the downside of our over-connectivity with our digital devices and highlights the benefits of real communication with the people we are with instead of checking to see what we might be missing online.

Building on this theme and from the same part of the world, a Brazillian bar in São Paolo called the Salve Jorge Bar has invented the "offline glass". This beer glass has been designed with a section cut out of its base, the exact size of a smartphone, so that it can only stand upright when it rests on top of a mobile phone. This glass has been designed to combat the texting epidemic prevalent in bars and restaurants, and according to them to "rescue people from the online world and bring them back to the real one". (http://vimeo.com/64643705)

This inspired design was created in collaboration with advertising agency Fischer & Friends, and was so carefully designed that it even addressed the issue of condensation, shaping the glass to avoid condensation dripping down on to the smartphone. The Salve Jorge Bar now only serves beer in the offline glass and even created a promotional video to explain the concept and show how the glass was made. Another extra bonus they may not have considered is that as your mobile phone is trapped by the glass it has the added advantage of preventing the sending of those drunk texts that invariably cause huge embarrassment the next day!

## In the bag

If you just want to take some time out from being plugged in but don't have the willpower to disconnect your smartphone on your own, you could always get a "Blokket", a phone pocket made from a smart material that blocks mobile phone signals. Once removed, your phone will resume normal service and receive any missed calls and texts.

The concept, designed by three Parsons School of Design graduates, grew from observing the domination of the smartphone in social settings, particularly in restaurants. Described as "A pouch that ensures you pay attention to your date, not your phone", it highlights the way in which we are allowing our digital devices to prevent us fully engaging with the people around us.

## In conversation

I recently met up with a friend in a lovely local café to catch up on each other's news. We sat on a huge leather sofa, ordered some green tea and a couple of not-so-healthy but beyond delicious accompaniments, and started talking. Halfway through the conversation I thought something was different about the café; it felt unfamiliar, and I couldn't quite pinpoint what it was. I looked around and realized it was the sound of people talking, of laughter, communication, making eye contact, touching. The café wasn't silent because no one had their head in a smartphone or laptop; everybody was talking! But when did this become the exception, when did this become something unusual?

Perhaps we've reached the point where we need new guidelines to show us how to balance our digital life with our real life; a new form of digital manners that prioritizes human connections over digital ones.

One of the best and hardest-hitting commentaries on how our digital obsession is ruining our relationships is "Touch" by Hollie McNish, a published UK poet and spoken word artist (see next page). Her poetry is rhythmic, deep and powerful, her subject matters delving into the issues of contemporary life. She is deeply committed to the idea of language as a tool of self-expression and communication.

She has released two critically acclaimed poetry albums, *Touch* and *Push Kick*, and a first collection of written poetry, *Papers*, published by Greenwich Exchange, London. She has performed at the Glastonbury festival, Ronnie Scott's Jazz Bar, London's Southbank Centre and Cambridge University and has had poems commissioned by Radio 4's *Woman's Hour*, WOW festival, Channel 4 *Random Acts* and The *Economist* Educational Foundation. Celebrated contemporary spoken word poet Benjamin Zephaniah said of Hollie McNish, "I can't take my ears off her." The YouTube video of her spoken word performance of *Touch* has had over 65,000 views. (https://www.youtube.com/watch?v=0lCG71LD8Sg)

### "Touch" by Hollie McNish

(edited excerpt)

We're sending fake arsed Facebook pokes and hugs

simulated sex and we're forgetting how to make real love

no more secret meetings under trees in springtime

just an x at the end of an MSN like that's a real sign

I want to see the unbuttoned shirts to knickers

circling chests with the tips of your fingers

I want hands on, lips on, heat on

eyes to eyes, a single biggest turn on

I want you to see me

I want you to feel me

I want to spend the day with you sweet day dreaming

PSYCHOBUBBLE

## Love needs five senses

*We are creatures of five powerful senses – smell, touch, hearing, taste and seeing. And yet the digital world privileges only one of these – the visual. This is leading to all kinds of problems in how we relate to other people, and ourselves.*

*When a baby is born, their ability to see is probably the least important sense they have. We develop strong feelings of love and safety with our caregivers through gentle touch, soothing voice, warm milk and familiar smells. This is the basis of a loving relationship and what is called "secure attachment". As adults, we also need to use all of our senses to make deep, connected relationships, and to feel love and be loved. As an agony aunt once replied to a woman who claimed to have fallen in love with someone online, "You can't fall in love until you have smelt someone!"*

*So it is important to take time to unplug from the digital, visual world and reconnect with the rest of our senses. Mindfulness gives us some very simple ways to start. When you have your first cup of tea or coffee of the day, take the time to really taste it. Bring the attention of your busy mind to the feeling of the hot liquid in your mouth, the texture on your tongue, the aromas of the steam. When you shower, wash your hands or brush your teeth, focus on the sensations, the sounds of the water rushing, the feeling of it on your skin.*

*By taking the opportunity to develop new habits of mindful awareness with these everyday tasks, we can begin to re-awaken all five of our senses and reconnect with the world, ourselves and others, and start to balance the visual tyranny of the digital world.*

**Jacqui Marson**

Chartered psychologist and author of *The Curse of Lovely*

## The innovation of loneliness

This statement inspired Shimi Cohen to create a powerful animated four-minute short film, which explores the correlation between our increased use of social networks and our augmented feelings of loneliness. The concept for the film was developed as part of Cohen's final project at Shenkar College of Engineering and Design. The film, which combines excerpts from Sherry Turkle's TED Talk, "Connected but Alone", and Dr Yair Amichai-Hamburger's Hebrew article called "The Invention of Being Lonely", examines the idea that we are using social networks as a substitute for the intimacy of personal relationships. (http://www.youtube.com/watch?v=c6Bkr_udado)

Online, we can present the pristine, controlled and curated versions of ourselves and find virtual popularity. We can be heard whenever we want to share something, often while it is happening, and can create the ideal digital world we want to inhabit, without the fear of rejection or being alone. We believe that being digitally connected automatically guarantees us a "delete" tab for loneliness.

But according to the research in the film, social connectivity does not guarantee that we will never feel isolated and alone. In fact it is exactly the opposite that has been found to be true. Our digital social life is the equivalent of a mirage in a virtual desert, creating the illusion of closeness and connectedness while in reality causing us to feel more alienated and isolated. Feelings of loneliness arise when we view digital connections as equal to our real offline relationships. If we eradicate all the possibilities to actually sit with ourselves and be alone, we are guaranteed to experience acute loneliness. The "Innovation of Loneliness" by Shimi Cohen has had over 5,000,000 views across YouTube and Vimeo.

**"Loneliness has become the most common ailment of the modern world."**

Shimi Cohen

## Out of con-texts

The telephone call is in decline and may soon be extinct, discarded in favour of our preferred method of shortfall communication, the text. The spoken word, spiked with emotion, interwoven with inflection, punctuated with silences and wrapped in unpredictability, is fast becoming a far too imperfect and uncontrollable method of communicating.

By contrast, texting, which uses a condensed, abridged version of language, does not require us to communicate face-to-face or directly listen to someone else's voice, and affords us total control over what we want to communicate. There are no awkward silences in a text, no human elements to distract us; instead we hide behind a reductive lexicon of words interspersed with the collection of cheery digital emoticons that we now use to symbolize our feelings.

Studies carried out by developmental psychologists researching the consequential effects of texting particularly on young people have unearthed concerns not only about their overuse of digital technology but about the negative impact texting can have on their interpersonal skills.

## Couples who text together don't stay together

The texting epidemic is nowhere more prevalent than in relationships. Couples now text each other throughout the day, at dinner, at the weekend, using it as a substitute for spoken conversations.

Apparently there is an interrelationship between perpetual texters and people who cheat in relationships, as lying in a text, without visible facial expression, eye contact and voice inflections, is infinitely easier and less detectable. Continuous texting may also impact on our ability to form healthy relationships in the future, as our skills in interpreting non-verbal signs become eroded through lack of practice.

As adults, we are increasingly using texting to hide, to avoid having conversations and to circumvent unnecessary dialogue. In relationships we can park the messy, painful, confrontational elements of any real, full-blooded adult relationship, and instead of working through the complex situations that enable a relationship to grow, we create an abbreviated, texted fantasy of perfect coupledom. But without sharing our authentic selves, are we any more than a couple of avatars trying to communicate?

Lori Schade and Jonathan Sandberg, researchers at Brigham Young University in Utah, conducted a study of the social effects of text messaging. Their research showed that an overuse of texting can actually disconnect couples as it removes the subtlety from a relationship. Their study surveyed 276 young adults across the US, 38% of whom were in a serious long-term relationship, 46% were engaged and 16% were married. The main points the study highlighted were that relying on texting as the primary form of communicating in a relationship only allows the expression of a limited, narrow and one-dimensional view. Although most of the couples surveyed texted each other multiple times each day, the study showed that the compressed vocabulary of texts does not allow for the expression of personality or emotion in the way that a real conversation does.

## "Selfie" improvement

Our obsession with needing to document every aspect of our lives via sharing, posting on Facebook, texting, tweeting, Snapchatting, Instagramming, Vining and Pinning seems to be fuelled by a mash-up of what previously, prior to the Internet generation, would have been considered the worst of our narcissistic and voyeuristic tendencies. Selfies are a prime example of how the self-promotional landscape has shifted. In the pre-smartphone days, the idea that we would take a series of pictures of ourselves (probably using a mirror) with a traditional camera, and then have the photos printed and mail them off to our friends saying "here are some great photos of me" would have been considered at best vain and at worst inappropriate or highly embarrassing. However, self-promotion in its digital format seems to have not only become totally acceptable but has become part of our everyday vernacular. It has even made it into the venerable Oxford English Dictionary in 2013 with the official title of "Word of the Year" and described as, "A photograph that one has taken of oneself, typically one taken with a smartphone or webcam and uploaded to a social media website."

**In 2012 *TIME* magazine undertook a survey that studied the way we use our digital devices. The study group consisted of nearly 5,000 people and spanned eight countries across the globe: the US, the UK, India, Brazil, China, South Korea, South Africa and Indonesia.**

**The survey also had some surprising findings that demonstrated how regional variances impact on our usage of digital devices, in particular the use of texts in and out of relationships:**

- 84% of people surveyed worldwide admitted that "they would not be able get through a single day without their mobile device"
- 80% of Americans aged between 25 and 29 admitted to taking their phones to bed with them
- 20% of people surveyed checked their phones every 10 minutes
- 25% of people admitted to checking their phones every 20 minutes
- 45% of South Africans said they have sent "sexually provocative" photos using texts
- 54% of Indians have sent provocative "sexts" accompanied by photos
- 64% of Brazilians sent texts and pictures containing sexual content
- 24% said they had used text messages to set up an encounter with someone with whom they were having an affair, which included in its findings 56% of the Chinese participants in the survey
- 25% of Americans admitted to sending sexually explicit content via texts
- 90% of Brazilians and Indians agreed that being constantly connected is "generally a good thing"
- 76% of Americans felt that being continuously digitally connected was a "positive" part of their lives

The selfie's definition has now expanded to encompass a picture taken of a group by one of its members. The most infamous one was taken during the 2014 Academy Awards ceremony by its host, Ellen DeGeneres, who in an inspired moment leapt into the audience and organized a selfie with a group of A-list celebrities. It fell to Bradley Cooper to hold the smartphone at the time.

DeGeneres tweeted in her inimitable style "If only Bradley's arm was longer," followed by "Best photo ever." This photo was viewed by an incredible 37 million people worldwide and became the most re-tweeted picture of all time, temporarily causing Twitter to crash.

The selfie syndrome has even permeated the hallowed inner sanctum of the Church and high government, with Pope Francis posting a selfie and First Lady Michelle Obama also sharing a selfie with the world. By contrast US President Barack Obama's selfie, taken with world leaders at Nelson Mandela's funeral ceremony, was internationally slated and deemed highly inappropriate.

The substantial increase in the popularity of selfies can be directly linked to the advances in the technological capabilities of smartphones, with the incorporation of a front camera function and integral instant editing facilities. As the quality of the inbuilt camera and lenses continues to improve in smartphones, we can expect an exponential growth in the taking of selfies in the future.

Beyond the selfie phenomenon lies a very basic human desire for validation and approval. We all like to feel appreciated, be liked and feel popular. Popularity is particularly important for teenagers, in particular young girls, a group within which there has been a marked increase in the curation of their own image by the posting of highly manipulated and retouched pictures of themselves in order to gain the maximum level of approval and positive feedback in the form of likes, shares and retweets from their friends. It has become one of the downsides of living life in the public domain and this trend has begun to show early warning signs for young teenage girls. It is developing into an obsession with potentially harmful real-life consequences.

**"Beyond the 'selfie' phenomenon lies a very basic human desire for validation and approval."**

Teenagers are increasingly "doctoring" their online profile photos and selfies using apps specifically developed to perfect their online image. These apps enable bodies to be slimmed down excessively, teeth to be whitened and straightened, and hair to be lengthened, and this has led to a marked increase in young girls seeking plastic surgery to bring their real-life image in line with their highly manipulated online images of themselves.

In order to restore some balance and perspective a new group of young celebrities like singer Lorde are starting to highlight photos of themselves that they feel have been manipulated by airbrushing or Photoshopping, and have instead posted pictures of themselves without makeup, setting the positive example that looking fresh and authentic is OK. This in turn has started a new positive trend, the "no makeup selfie", which provides a valuable counterpoint to the trend of posting highly manipulated images of ourselves. It also encourages women to value and share their natural beauty and build their self-confidence while promoting the concept of natural beauty. It has been adopted by charities to raise funds too. The most prominent charity using selfies in this way is Cancer Research, which raised $3.3 million in donations in 48 hours using the Twitter hashtag #nomakeupselfie. ActionAid is another charity that has adopted the selfie as a fundraising tool, with their launch of the "Selfless Selfie" to raise awareness and funds for their relief work in the Philippines.

Aside from its valuable fundraising capacities, the taking of these instant self-portraits has been turned by social media intern Robbie Jones into an art form with his "Selfie NYC" project, a Twitter-based callout for people in New York to post their selfies on the project wall.

For teenagers, as with all digital connectivity, it's all about balance. Selfies, if used properly in an authentic and un-manipulated way, can provide a creative outlet for young people to share who they are and what they are doing with their friends. It is a medium that, used in moderation, can enable them to visually express themselves, share their personalities and provide a chronicle of their lives through a series of selfie-generated "vision bites". However, the larger question remains as to whether selfies have a positive or negative impact on our self-confidence and the way we value ourselves. Is there a deeper message behind the placing of such importance on our appearance and the way we share photos of ourselves, so that ultimately that becomes our most important feature and the most valuable part of ourselves in social media terms?

### Selfie stats

**There are over one million selfies taken each day**

- 36% have been digitally enhanced in some way
- 34% of men admit to digitally enhancing every selfie prior to posting
- 16% of women admit to retouching every single selfie before it is posted
- Selfies now make up 30% of all photographs taken by 18–34-year-olds

## Food matters

Instagram has a lot to answer for. The photo app, which was launched in 2010, has fundamentally changed the way we eat. Gone are the days when we used to enjoy the anticipation of the flavour and aromas of the food, whether prepared by us or by someone else; those times when we would take time out to really connect with the people with whom we were sharing a meal, when we used to engage in conversation, share news and generally enjoy the simple pleasure of eating. Since the addition of Instagram to our digital lives, we seem to have become incapable of eating without first taking a photograph. Instead of enjoying a dish, we point, shoot and post. This obsession with documenting every aspect of our lives for public consumption is now also ruining our dining experiences.

Our obsession with "foodspotting" has reached epidemic proportions. While the presentation of food clearly has an important role to play, most of the food we eat or cook or that is prepared for us is not photogenic or created with a photo opportunity in mind. Food is about the flavour. We can't post a picture of the way food tastes. Ultimately it is the taste, the texture and the play of flavourswhich drives the food experience, none of which can be shared in a photo.

Restaurants, aware of the disruptive element caused by our addiction to photographing our food, have started to take action by enforcing restrictions on customers preventing them from doing this. In New York, several restaurants in Manhattan have implemented camera bans, including David Chang's celebrated upmarket restaurant Momofuku Ko.

Others, rather than implementing a total ban on photography, have taken a more lateral approach. In both of his restaurants in Manhattan, chef David Bouley decided that rather than allowing photography in his restaurant, he would offer his customers the opportunity to come into the kitchen and take pictures of their food. "This prevents flash photography destroying the atmosphere and disturbing other diners while they are eating in the restaurant," he explains. Moe Issa, the owner of Chef's Table at Brooklyn Fare, found another solution. Having originally banned photography in his restaurant, staff will now email photos of the food customers have eaten directly to them the following day.

New York eating house Comodo decided to buck the trend for banning photography by actively working our food photography obsession to its advantage. Instead of discouraging customers from taking pictures of their food, it asks them to photograph their plates of food in order to create a searchable, photo-led menu. Each menu then has the relevant hashtag added to it so that diners can see how the plates will look before they order them.

In the UK, our food photography fixation has been used as a bartering tool by one food manufacturing giant, Birds Eye (of fish finger fame) to provide free meals in exchange for customers posting pictures of their food on Instagram and Twitter.

To launch the campaign, a series of independent pop-up restaurants, aptly named "The Picture House", were set up in various locations. Diners who posted photos of their food would receive a free meal. Presumably, the idea behind giving away meals in exchange for social media posts is that it will provide an initial surge in free publicity, but is this a viable long-term option for a food brand whose business is based on the selling of its food?

**Our widespread addiction to sharing food photos does not seem to be limited to one type of food, as shown by the survey conducted by Mashable.com in 2011 on the most popular foods to be photographed:**

- 72% of the photos taken of food were of a main meal, with dinner being the most posted and shared meal
- 25% of people, when asked why they shared photographs of their food, had no specific reason other than it "was just what they ate that day"
- 18.3% took photos of desserts, which was the most popular subject matter
- 17.8% snapped away at vegetables
- 13% took photos of poultry
- 10.7% photoblogged meat pictures
- 8.8% took pictures of bread
- 7.8% took photos of drinks
- 7.1% photographed dairy foods
- 7% photographed pasta (the least popular category)

## Lost your appetite?

Our addiction to continuously taking pictures of what we eat, aside from upsetting our fellow diners, may actually be ruining our appetites.

According to a study by Brigham Young University, based on 232 people who were asked to look at a large selection of photographs of food and rate them, "Over-exposure to food imagery increases people's satiation." They define "satiation" as the drop in enjoyment we experience with repeated consumption. In other words, looking at too many photos of food can actually have a negative impact on our capacity to enjoy our food, because we feel as if we've already had the experience of eating.

Part of the pleasure of eating is anticipation. By saturating ourselves in food imagery, we lose the opportunity to look forward to what we are about to eat. Co-author of the study Professor Ryan Elder describes it as "sensory boredom – you don't want that taste experience anymore."

During the study, half of the sample group were shown 60 photos of sweet foods like cakes, chocolates and pastries, while the rest of the group were shown 60 pictures of salted foods such as French fries, crisps and pretzels. Each group was then asked to rate each picture based on how appealing that food appeared to them. At the end of the trial all the members of the study group were asked to eat some salty peanuts and rate their level of enjoyment. The findings revealed that the people who had been shown photos of salty foods ended up enjoying eating the peanuts less, even though they hadn't looked at pictures of peanuts and were only shown pictures of other salty foods.

## Guilty pleasures

Life is full of basic pleasures, such as our first cup of coffee in the morning (green tea in my case), eating a really well-cooked meal (without Instagramming it), watching a great movie or even having sex, so the idea of having to choose between one of those essential pleasures and a mobile phone can be at best challenging, at worst nearly impossible.

**By saturating ourselves in food imagery, we lose the opportunity to look forward to what we are about to eat**

As part of a study conducted by MobileInsurance.co.uk in the UK, a study group of 2,570 adults aged between the ages of 18 and 30 were asked, "What would they rather live without than their mobile phone?"

Some of the findings were surprising, if not inexplicable, such as the 66% of the study group who claimed that they "couldn't live without their smartphone". Incredibly 9% said that they would prefer to part with their own children than be without their digital devices!

The most surprising finding of the survey has to be that 94% of adults aged between 18 and 30 in the UK would prefer to live without sex than their mobile phone. This is either a dire commentary on how much we Brits like sex, or on how much more we love our mobile phones. Either way it is an alarming statistic.

## Love at first site

For centuries we have sought assistance in meeting our ideal match. We have explored a variety of ways to find "the one" – from using matchmakers to fortune tellers predicting we'll "meet a tall dark stranger", through to blind dates and singles nights, we have always searched out ways to meet our life partner.

With the dawn of the Internet age in the mid-1990s came another alternative. The world of online dating revolutionized the way singles met by providing a new way to find a partner with a global reach. Sites such as match.com and eHarmony were the early pioneers of online dating. Over the past decade these sites have developed the way in which they operate and can generally be split into two categories.

Some sites have a more "hands off" approach, in which users create their own profiles and are encouraged to browse others' profiles on their own. Other sites participate more actively in the matchmaking process, using questionnaires and computer algorithms to match their subscribers.

There are currently over 40 million people using online dating platforms to find a partner. Gone are the days when online dating had the stigma of being a potentially risky way to meet people. The arrival of the smartphone has heralded an even more portable version – the mobile phone dating app. According to the Pew Research Center's study into whether online dating actually produces more successful relationships than offline dating, the results are inconclusive. Clearly online dating does offer opportunities to meet people outside one's direct social circle, and provides a filtering opportunity prior to meeting face-to-face. For others it is often the only way for people who may have more limited options thanks to age or geography to actually meet people.

Online dating has had an impact on the number of young adults deciding to get married – but not in the way we would expect. One of the consequences of having a permanently available and receptive online pool of potential partners has been to reduce the number of, in particular, men, who are committing to getting married. This form of online "love supermarket" has created less of a need for long-term relationships, as replacement partners are permanently available on tap.

Some sites have actually used the growing demand for casual encounters by providing mobile dating apps that take the "instant gratification" element of online dating to another level. Often based around location using GPS, these apps can instantly connect people with nearby matches. As mobile dating apps enable instant real-life meet-ups, it is important for users to remember to implement some safety precautions. Staying in public places, communicating via the app instead of sharing personal contact information and creating a special username rather than using your own name will help to ensure that these casual meetings are undertaken safely, particularly as they tend to happen more spontaneously

### The appy couple

As an alternative to sharing relationships publicly on social media networks, a new breed of social apps have been developed to provide private online space for couples to share messages and posts.

These apps can connect couples on a variety of levels. Apart from providing a private platform for them to communicate, couples can share intimate messages, photos and even their to-do lists in complete privacy. Relationship apps are also beneficial to couples in long-distance relationships or those who have to be apart for long periods of time, as they provide real-time connection in a way that otherwise would not be possible.

**"Social media can provide a tempting platform for avenging our newly 'insignificant' other."**

## Revenge posting

Not all relationships work, whether they begin online or offline. The difference now is that once a relationship has failed, depending on the circumstances of its failure, social media can provide a tempting platform for avenging our newly "insignificant" other.

Social networking offers an overwhelming array of options for monitoring and checking up on newly-ex partners, and also provides the perfect forum for posting photos and updates about a new date. The practice of following an ex online and posting photos that say "I'm fine without you" is all too common among 18–29 year olds. Over 30% of young adults have admitted to either monitoring the posts, photos and relationship status of an ex-partner or posting photos and status updates specifically designed to make their ex jealous.

## I'll have an "old fashioned"

Although for over 40 million people online dating sites are the go-to option for meeting people, a psychological study commissioned by the Association of Psychological Science says that we may actually have more success meeting a future partner the old-fashioned way, in an offline social setting such as a bar or party.

The study found that the algorithms used by dating sites are not able to do much more than compare people's characteristics and interests and match them based on that information. According to Eli Finkel, associate professor at Northwestern University, these algorithms cannot predict whether there will be chemistry and attraction between two matched people even if they have shared interests.

Finkel believes that there is no better way to establish whether two people are compatible than by meeting in person and "having a cup of coffee or meeting for a drink". The study found that there were several further negatives to using online dating to find a partner. Although online dating sites provided access to a huge volume of potential partners, the enormous amount of prospective matches can be overwhelming and lead us to undergo a form of "shutdown". This in turn can drive people to make misjudged choices through having too many options available to them.

Interestingly, Finkel cites science as the

reason algorithms cannot successfully match two people. He explains that in 80 years of scientific research into relationships, purely using background information to match people cannot correctly predict whether a relationship will be successful. Having reviewed all the findings, Professor Finkel concluded that contrary to the popular assumption that online dating sites work, he was of the opinion that they actually do not work.

## The swinging 60s

One of the sectors with the greatest uptake of online dating is the more mature market. Over 50% of adults online in the US are over the age of 65. Adults in the 55–64 age bracket are also the most rapidly growing sector on Twitter, with 45–54-year-olds showing the most rapid growth on both Facebook and Google+. According to a report by the Pew Research Center, over 70% of seniors are actively going online every day, although Internet usage mostly reduces after the age of 75.

One of the increasingly popular uses of the Internet by the older generation is for online dating. According to statistics from leading online dating site match.com, their fastest-growing sector is the 50+ age group, which has grown by 89% over the past five years.

Helen Fisher, a consultant to match.com, is not surprised by the uptake of online dating by seniors. She believes that older people are looking for a different type of relationship, such as a companion or a lifelong partner. In her experience, the older generation are also much more open to meeting people from a different background.

Unlike younger singles, who often use online dating sites to pursue more casual relationships, the older generation are mostly searching for meaningful, long-term relationships. The desire to find a "significant other", coupled with the increase in later-life divorces and a general lack of social opportunities for older adults to meet new people, has increasingly driven the post-50 generation to move online in their quest to find new prospective partners.

One of the benefits of mature dating is that at that stage in their lives older daters are much more likely to have a clearly defined picture of what they are looking for in a potential partner, and will often have a much greater level of wisdom and experience born of past relationships.

However, many senior daters, who are perhaps taking their first steps into the world of online dating, feel intimidated by the huge selection of online dating sites available to them. Once they have signed up to one or more sites, they can often feel overwhelmed by the idea of creating a profile and sharing personal information.

## Catfishing

Research has shown that mild forms of misrepresentation online are prevalent everywhere, but they are mainly born of our desire to market ourselves online to our best advantage. Mostly harmless, they manifest themselves in amended vital statistics, such as men stating that they are taller than they are or women presenting themselves as thinner or younger. These subtle distortions of the truth are not harmful or intentionally designed to hurt anyone – but the practice of creating a false identity can have a very different effect.

When it comes to online dating, being deceived or duped into having a relationship with someone who misrepresents themself is one of our greatest concerns. This form of online deception has a name: "catfishing". According to the urbandictionary.com, catfishing is defined as "The phenomenon of internet predators that fabricate online identities and entire social circles to trick people into emotional/romantic relationships (over a long period of time)."

Since its original definition, the term has expanded beyond people intentionally deceiving a potential date into a relationship to include anyone who misrepresents themselves online.

The term "catfish" was adopted by Nev Shulman, a filmmaker who himself unknowingly became a victim of online duping in his personal life. This inspired him to make a documentary film about his experience, which he called *Catfish*. The film attracted so much media attention that it led to a large number of people contacting Shulman to tell their own stories of being deceived into having relationships online. These personal stories led Schulman, with co-filmmaker Max Joseph, to create the cult MTV series of the same name. In the series they help people who have never met their "partner" in real life, only online, find out if the person they believe they are in a relationship with is lying about their identity.

Having watched the series, there seem to be several predominant characteristics shared by the "catfishers", which form a continuous thread throughout the episodes. Although there are always exceptions, a large number of the people who create fake online profiles are disadvantaged in some way. Many of those featured in the series were lonely and had very few offline friends. Others were very overweight and clearly found making friends in real life challenging, so they sought to create the sort of online image that they couldn't hope to have offline.

Geography also had a significant role to play, as many catfishers lived in remote and isolated locations with very little opportunity for meeting people. Very rarely was the motive malice or knowingly wanting to hurt someone. Mostly, people just wanted to have the sort of idealized life and relationship online that they would have struggled to find offline. The grey and complex area of having a relationship with someone with a fake profile is essentially that their unknowing online partners had fallen in love with an image that wasn't real but, more often than not, a personality that was.

# 4 PAUSE/ DISCONNECT

## Live smart: in search of digital underload

**It may seem obvious, but we actually do have the option to turn down the volume on our insistent world. We can reject the dehumanizing, desensitizing effects of digital overload and "choose life" every single day. In order to be able to take a step back and think about the way we use our digital devices – and the way they are both using and losing us – we need to disconnect. We can decide to change the way we live and take those first steps towards a life that enables us to be the best version of ourselves. Every day, we can do one thing that will take us forward to where we want to be. And there is another upside: when we decide to take the first step and let go of our need to be digitally connected 24/7, that need also lets go of us.**

Instead of allowing technology to control our lives, we can learn to access our "inner technology" and find new ways to live deliberately, consciously and mindfully. Unplugging enables us to recharge the mind, body and soul.

In order to generate some space between ourselves and our digital devices, we need to create zones in our life that are unplugged. Once we can step back from the barrage of excessive information for long enough to allow us to find presence, we can then ask the questions that can only be answered in a space free from digital overload. Digital downtime enables us to integrate the different spaces in our lives and start to live life to its fullest potential. There are different methods that we can use to disconnect from the digital noise with which we surround ourselves and that will enable us to find new ways to look within ourselves and reconnect with the essence of who we are.

Three of the most effective techniques are meditation, yoga and mindfulness (see also chapter 5). Here is a brief introduction to each of these techniques:

### Meditation

There are many myths surrounding meditation, which is often viewed as a somewhat elusive and mystical practice. In fact it is one of the most elemental things that we can do because it centres around our breath, and breathing is something we all do. Meditation doesn't require any special training or equipment, as we already have everything we need within us.

Meditation is essentially the process of clearing the mind, focusing on the present moment and letting go of all the mental clutter in our heads. With meditation there is no right or wrong way; it is about finding the place in you that "is" and from there reconnecting with yourself. According to Zen meditation master Thich Nhat Hanh, a global spiritual leader, poet and peace activist, "The breath is the intersection of the body and mind."

There are different types of meditation practice, so it is worth trying out a variety to see which one you connect with the most. By spending just 5 minutes a day meditating you will find your capacity to concentrate and problem solve greatly increased. It has a profound impact on our physical and spiritual wellness, and will provide a gentle respite from the incessant distractions of our digital world.

Meditation is not a "quick fix" practice. It is a slow, continuous, self-motivated discipline to enable us to reconnect with our inner being. It is a practice that encourages slow, steady growth. It is important to understand that the

**In order to start meditating there are six basic things you can do to prepare yourself that apply whichever type of meditation practice you decide to follow.**

- **Make a sanctuary space.** Find a quiet place that you connect with, which has few distractions and where you will not be interrupted. This could be a special corner of a room, a bench in the garden, a comfortable armchair or a favourite rug, anywhere that can become your special place, where you can find a peaceful "pause".
- **Find your position.** Try out a few seated positions until you find the one you are most comfortable in and are able to remain in for the duration of your meditation. Meditation is mostly practised in a seated position to avoid the risk of falling asleep, but depending on the type of meditation you can practise lying down, standing and or even walking.
- **Shhhhhh....** Prior to meditating, in order to bring a quietness to your mind, it can be helpful to sit in silence, focusing on your breath until your heartbeat slows down.
- **Centre your focus.** Focusing the attention forms an intrinsic part of meditation practice. Usually this will be on your breath, on a mantra (a phrase or sequence of words) or on a specific object. In other types of meditation you will concentrate on whatever thoughts dominate your consciousness.
- **Let it be.** In meditation it is conducive to try to view your mind as an "impartial observer" by letting thoughts and distractions come and go naturally without judging them. When we find our attention wandering, rather than overriding them by trying to suppress them, we gently bring our attention back to the centre of our focus, objectively observing our thoughts and emotions.
- **Small steps.** At the beginning of your meditation journey, it is important to remember meditation is a process; it does not have a destination, so the setting of personal challenges such as meditating for a specific length of time or reaching the "peace spot" are a complete antithesis to the essence of meditation. As meditation becomes more integrated with your daily life, the length of practice will gradually increase naturally and without requiring conscious effort. Over time you will find that you can meditate whenever you wish and wherever you are.

positive effects and changes of regular meditation practice are subtle and may not be immediately noticeable. Over time these changes will become more apparent in our improved ability to handle stressful situations calmly, to focus on tasks fully and to live life more mindfully.

**"Silence, I discover, is something you can actually hear."**

Haruki Murakami

For many, the practice of meditation acts as a powerful counterpoint to our digitally overloaded lives. With its roots deeply entrenched in ancient spiritual traditions it seems paradoxical that the digital world that we are seeking to escape could also provide us with the tools to enable us to meditate more effectively. Available in a multitude of formats from apps to online guided meditations and podcasts, the online world of meditation provides every type of meditation from a quick on-the-go "3 minute meditation" to "soothing anti- stress meditations" and even a "chocolate" meditation!

## Yoga

Yoga impacts every aspect of our lives. There are over 20 different types of yoga and several more branches that combine different yoga techniques. Within this broad range of different styles, yoga offers several techniques that can be perfectly matched to and effectively treat the different conditions that arise from digital overload. Yoga is a holistic practice that integrates mediation, breathing and movement to harmonize all the elements that make us who we are.

Internationally acclaimed Yoga teacher Lisa Sanfilippo teaches yoga in London, throughout the UK and internationally. She established her dedicated yoga practice in 1997. Her teaching incorporates Tantric philosophy, transpersonal psychology and yoga therapeutics.

**"For a holistic form of yoga that improves our daily lives, I combine a Tantric philosophy of intrinsic goodness, good physical alignment and an understanding of the psyche. This helps us to unplug from technology and plug in to what really matters in a clear cohesive way."**

Lisa's yoga technique is "a fusion of the spiritual and the physical" and is designed to keep these in balance with the "virtual". Lisa believes that in order to detach from information overload we often need to "reset" our internal operating systems so that we can reconnect with the natural rhythm of life and rediscover how to live "in the moment". Her approach to yoga reconciles our personal experiences with our universal experiences of being human and is designed to wake up the different parts of the body. Lisa likens the body to a home and her mission is "to make sure all the lights are on in every room".

Lisa's approach to tackling information overload starts with the breath. Breathing exercises are the foundation of yoga. As human beings we are a multi-layered fusion of physicality, energy, thought, emotion and wisdom, and when some of the layers are blocked there is an imbalance that can lead to ill health or crisis. Lisa's yoga practice works to harmonize all the elements contained within us and bring us back to a position of wellness and positivity. She works from the inside out and the outside in: her aim is to get us to move enough to be clear and use our energy to get to a place of stillness and clarity, helping us to become more focused and effective.

Lisa believes: "When we connect the body, the breath, the mind and the heart, we can focus on what really matters."

**Yoga Snack**

### Balancing Breath

Increase your focus by calming and balancing the nervous system, decreasing mental tension, and connecting to your physical and mental centre. Breathing with an even inhalation and exhalation balances the activating and relaxing parts of the autonomic nervous system. This basic yoga breath is subtle enough to do anywhere, any time, when you are feeling stressed or tense, or have a buzzing brain.

- To learn this, start by making a whispery "haaah" sound – breathing out with a gentle constriction at the back of the throat.
- Place your palm in front of your mouth and pretend it's a mirror; use your breath as though you're going to fog it up.
- Next make the same "haaah" that tones the back of the throat as you breathe out with your mouth closed.
- Breathe through your nose keeping the tone in the back of your throat. Now try the tone in your throat as you inhale.
- Continue inhaling and exhaling with that tone in the throat that lengthens your breath. Breathe deeply and comfortably into your lungs, as though filling the bottom back part of your lungs with air as you breathe.
- Your inhale and exhale should be the same length by counting: inhale-two-three then a gentle pause, then exhale-two-three. It may take some practice to get your breath to feel fluid, without any pauses or jagged stops during the in-breath or out-breath.

If you find the breath "catches", this is a sign to soften your throat so you can breathe smoothly into your lungs. It's important to focus on filling the whole of your lungs, and keep the tone in the throat very soft, so that it is just a marker of the breath and doesn't create any additional tension.

**Lisa Sanfilippo**, yoga therapist and teacher

### Mindfulness

Mindfulness is a way of living. Being mindful changes our experience of the world, and it has been scientifically proven that mindfulness can have a positive effect on our health and emotional wellbeing. As we deal with an increasingly complicated and changeable world, practicing mindfulness and fully focusing on one thing at a time increases our capacity to handle complex situations with ease and clarity (see pages 82–83).

## Digital detoxing

The term "digital detox" has officially made it in to the dictionary, albeit (somewhat ironically) only in the online version of the Oxford Dictionary.

A digital detox is described there as: A period of time during which a person refrains from using electronic devices such as smartphones or computers, regarded as an opportunity to reduce stress or focus on social interaction in the physical world: break free of your devices and go on a digital detox.

Unplugging is not just about rejecting or parking our digital devices. It is about being in a space that allows us to recharge and reconnect with the natural rhythm of life and gain new perspective on our relationship with ourselves and with the technology we use. It is about re-evaluating our priorities, which in turn will enable us to establish a new set of core values to live by and find a more mindful and meaningful way to live – both offline and online.

Undertaking a digital detox by unplugging has been shown to substantially reduce the symptoms associated with digital overload and tech addiction, such as stress, anxiety, exhaustion and depression. Digital detox programmes are specifically designed to explore the benefits of unplugging on our mental, physical and social wellbeing.

If the idea of "going it alone" on a digital detox feels too challenging and overwhelming, the rise in the demand for tech-free tourism means that there are an increasing number of spas, wellness retreats and specially created digital detox camps that will guide you through the process. (See the descriptions of the top digital detox retreats on pages 120–125).

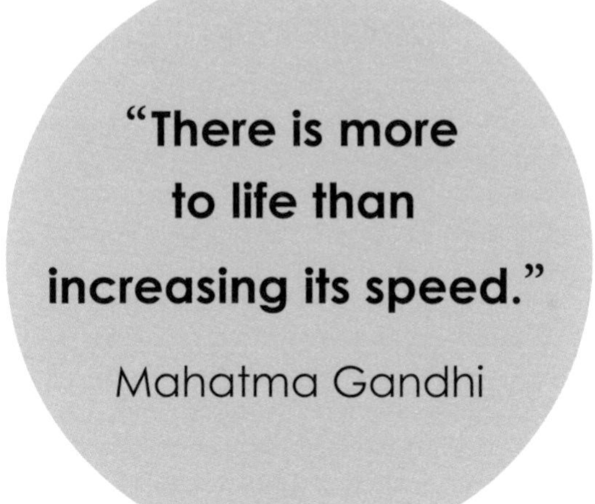


### The art of slow

The art of slow is about creating space without the need to fill it. A space that will allow you to "be" and listen to your inner voice, and rediscover your authentic self. Taking a step back by putting your life on pause will encourage self-reflection and enable you to reconnect with yourself and re-establish meaningful relationships with those around you.

The term "slow" makes us automatically think of a speed, but it is not really referring to the rate at which we do something. The "slow" philosophy is much more about finding a way of doing everything at its own natural pace, in the best way that you can. It focuses on the quality rather than quantity of what you are doing.

**"Slow isn't necessarily a time frame, it is more an ideology, a perspective, a more connected view of life."**

In a way digital detoxing is a form of going slow. It is taking a pause to enable you to find real connection without the permanent pull of digital devices. People used to take time to meet and talk to each other face-to-face or on the telephone. Communities were built on the communal activities and the shared resources of its residents: growing things, cooking, local sports – all social events. Today many of those simpler, real, experiences have been replaced by their virtual counterparts, which occupy our time and yet leave us unfulfilled because, ultimately, our digital devices cannot express emotion or give us a hug.

Shifting gears to a slower, more considered pace enables you to become more focused on the "micro" moments and to savour each detail – the sort of moments that we miss when we are permanently running to catch up with the demands of our cyber lives.

Although the world we live in is digital-based, we need to remember that everything in the world is not digital. While the advances in technology have enabled us to do things unimaginable 20 years ago they have also thrust our lives into fast forward. We are permanently overscheduled and overstretched; for many of us rushing has become the normal pace for all activities, both at home and at work. We take no time to prepare food and even less time to eat it. We do not read any more, we just skim headlines in 140-character soundbites. We no longer enjoy the journey as we "just need to get to where we are going". We can give ourselves a respite from the manic pace of our digital lives by giving ourselves a breathing space that will enable us to acknowledge the disconnect that digital over-connection creates, which can contribute to feelings of anxiousness and stress. "Pausing" and stepping back from the white noise of digital overload and excessive media consumption will enable us to reconnect with the natural rhythm of life and re-learn how to live mindfully in the moment. Embracing the "art of slow" will enable the mind and our emotional inner life to be able to breathe and recharge. This will help us to gain a new perspective on our relationship with ourselves and the physical and natural world around us, giving us the tools with which to re-evaluate our relationship with digital technology and find a more mindful and meaningful way to live and re-establish a balance between our online and our offline lives.

## 16 simple ways to pause and embrace the "art of slow"

1 Be silent. Give your mind some space.

2 Breathe consciously (see Yoga Snack, page 79).

3 Find time to meditate.

4 Spend some quiet time alone.

5 Write down your thoughts and ideas.

6 Draw, paint or make something.

7 Sing or make music, alone or in a group.

8 Walk everywhere, as often as you can. Instead of getting a drive-by coffee from a multinational chain, walk to your nearest independent café and build a relationship with your barrista.

9 Discover what your neighbourhood has to offer by walking or cycling there and explore the richness and variety of local shops and markets.

10 Smile at a stranger. It doesn't cost you anything but that moment of connection may just make their day and yours!

11 Talk to others.

12 Go to a park you have always driven past but have never stopped at.

**13** Immerse yourself in nature and wildlife. Visit a bird sanctuary, national park or hiking trail.

**14** Source locally grown produce at local farmers' markets instead of pushing a metal trolley around a brightly lit supermaket. This supports your local community and the environment and goes a long way to helping you reconnect with your neighbourhood.

**15** Find out about the provenance and backstory of your "farm to table" food. This will establish a connection between you and the grower and fundamentally change the way you experience the eating of that produce.

**16** Grow something from seed. The process of planting a seed and watching it grow through constant care and nurturing is the perfect analogy for living a slower, more connected and authentic life. We cannot hurry nature – everything happens in its own time. In the same way, if we connect with the natural rhythms of our world and respect them rather than trying to manipulate and control them to fit in with our digitally determined schedules, we will find that we do not have to stop moving in order to slow down; we just have to relearn how to move in a different, and more meaningful, way. We simply have to be able to pause for long enough to notice the details that shift our lives from the manic to the micro.

5 RECONNECT

## Taking it back to simple in an over-complicated world

**Our world is complex, overloaded and overwhelming. Increasingly we are finding that we need to step back from the tsunami of excessive information that being permanently online creates. We need to pause for long enough in a space free from digital overload to be able to find presence and be able to reconnect with the inherent rhythm of life. This will gradually enable us to find the place in ourselves where we are, and give ourselves permission just to "be".**

**"We are human beings, not human doings."**

Wayne Dwyer

We are amazingly wonderful, flawed, individual, imperfect human beings, and that is what makes each of us unique. We are not meant to be everything, do everything and know everything, especially not at the same time. Our imperfections and vulnerabilities as well as our talents and capabilities are what define us. It's OK not to be perfect. In Japan they actually celebrate the beauty of imperfection within their philosophy of Wabi Sabi, which is often expressed by the idea of filling in the cracks of a broken vessel with gold.

**"Accept 'what is' and 'who you are'. Imperfections are what make us human."**

I consulted with mindfulness and digital detox expert Dr Barbara Mariposa about the unique technique she has developed to combat the stress of digital overload. Called MMM™, or Mind Mood Mastery, it combines her traditional medical background with wellbeing psychology, neuroscience and mindfulness practices.

**"Being well is much more than just the absence of illness."**

Dr Barbara Mariposa

Her technique provides powerful tools to get well and stay well. She believes that "truly successful people don't have a problem investing in themselves, being honest about where they are in their lives and in themselves, recognizing the danger signals and listening to them. They know that investing in their health is always going to be a sound investment. To be truly well and function effectively, both our emotional and our mental wellbeing need to be working in harmony". According to Dr Mariposa, "This is where it all starts: 'The Inner game of Health'. Being well is much more than just the absence of illness. The aim is to flourish fully with a sense of being comfortable in one's own skin, living a life that is the right fit for you."

## Digital overload: a process addiction

Dr Mariposa explains the science behind the medical impact of digital overload on our body in this way: when we are suffering from "busy brain syndrome", with multitasking being a prime example, our body goes into a "a state of stress" and responds physiologically with the adrenal glands releasing more cortisol. This in turn has been shown to interfere with our ability to assimilate information, as well as lowering our immune function and bone density, and it can lead to several chronic conditions such as increased weight gain, high blood pressure, high cholesterol and heart disease. The role of cortisol within the body is as a substance to be released by the adrenal glands in response to stress, as a part of the fight-or-flight mechanism. This physical response is essentially designed to be used in an emergency.

Living our digitally overloaded lives on turbo charge has become our default setting. It is the equivalent of having the accelerator pedal on our adrenal glands, set at full throttle, releasing continuous amounts of cortisol into our system. This is equivalent to running on empty with no reserve battery.

We are fuelled by an underlying fear that if we were to actually unplug, we would be rendered ineffective and unproductive – or worse, actually fall apart. In fact, according to research, the opposite is actually true. According to Dr Mariposa, we have "insufficient bandwidth" to cope with the levels of digital overload that we are experiencing. Our bodies and brains can't cope with the excess amounts of cortisol or the increased speed of beta waves in the brain that occur when stressed, rather than the alpha waves that occur when we practise mindfulness. When we are less active, focusing on one thing at a time – unitasking rather than multitasking – we become more connected and productive and are able to respond appropriately to the situations we face.

**"Living our digitally overloaded lives on turbo charge has become our default setting."**

According to scientific research, the human brain can only perform two separate tasks simultaneously. When the brain has to undertake two different tasks at the same time, the prefrontal cortex has to divide into two halves so that each half can focus on one of the tasks.

The anterior part of the frontal lobes is the area in the brain that enables us to switch between two tasks. When an additional task is added, it has been shown that, although the brain still has the capacity to deal with this, concentration and productivity decrease in doing so.

Further research studies indicate that multitasking leads to less effective filtering and retention of important information. We experience a diminished ability to assimilate data outside of the task-related information, and difficulty with recalling information from both our long and short-term memory, making it more difficult to switch from one task to another.

## We are not our thoughts

Dr Mariposa explains that just 10 minutes in a state of calm renders an elastic quality to our subsequent experience of time. Time seems to expand as we unplug digitally and "plug in" to ourselves. As we focus on our experiential selves we learn to "walk alongside" our thoughts and feelings. We are not our thoughts but rather a vessel to hold our narrative self. When we are living lost in thought, physically we are responding on autopilot.

## Digital clutter

Mindfulness actively enables us to access the space within ourselves between stimulus and response, disabling our "digital clutter" and achieving a state of mindful calm. Through mindfulness, the aim is to free our thoughts and remove the power they have over us. Acceptance of what is and who we are has a key role to play in the development of our emotional intelligence. It is the counterpoint to the permanent state of stress and tension that we create by "trying not to fall". Learning that it is OK to be imperfect, to fall, to fail, can act as a potent catalyst for self-expression and ultimately self-actualization.

# Changing your "to do" list for a "to be" list: 8 mindful steps to bring you back to yourself

Practising mindfulness can help you to access the pathway to your instincts, your intuition and your wisdom. These simple techniques encourage us to be aware, conscious and mindful, so that we can rise above the noise and re-learn how to spend time with ourselves.

## 1 Breathe

We breathe unconsciously all the time. Even if we don't consciously make the effort to breathe, we still breathe. So if we make a conscious effort to concentrate fully on our breath, it is one of the purest, most balancing and life-enhancing things we can do and is fundamental to our wellbeing. So whenever you can, take a few minutes away from your digital life and allow stillness in. Feel your breath. Focus on breathing in slowly and breathing out even more slowly. Breathe consciously. Take three deep breaths every hour throughout the day. Embrace the rhythm of your breath like the sound of the wind or of gentle waves. Find your special space; your inner happy place that is always there with every breath.

## 2 Look up

We can't experience the light and dark of life through a 2-inch backlit screen. Remember that above us is a huge, expansive sky. Ever changing. Always there. Looking up gives us context and helps us to regain a more balanced perspective and remember to explore the outer edges of life. Even if you can only take 5 minutes to look at the sky through a window, take those 5 minutes for you. Watch the clouds, focus fully on the changing shapes, look at the shadows they cast. Be present. Those 5 minutes are always there for you take on a "need to slow" basis

## 3 Allow the "now"

Simone Weil wrote: "Attention is the rarest and purest form of generosity." Now is all there is. When we fully concentrate on being in the "now", we are free. Free of "what if", "I should have" and "I wish I could". They don't exist. Yesterday doesn't exist any more. Tomorrow hasn't arrived. All we have is now. So if we can focus fully on this moment right now, it is the purest form of being. It simplifies everything. Whether you are doing something, going somewhere or talking to someone, be there, be present, be at one with yourself and be secure in the knowledge that where you are right now is exactly where you are meant to be.

## 4 Get out more. See a tree

Think outside. No box required. Being immersed in our digitally connected world mostly means that we are living inside our heads and inside buildings. Our bodies and our senses are starved of attention and are suffering from abandonment anxiety. It is so easy to forget that the body directs traffic. It is where everything is housed; it is our very own mainframe computer. Yet we seem to have to continuously remind ourselves to physically reconnect with our bodies. We just have to get out more. Recent studies have shown that we need to change our level of activity every 90 minutes. Sitting for extended periods has been shown to be seriously detrimental to your health. Even if it is only for a 10 -minute break – leave your smartphone behind, take a walk, focus fully on what you see. This will encourage a new dialogue between the brain and the body, allowing the mind to relax and decompress enough to give it the space to connect with our emotions. Even without having an overtly natural environment around you, the act of walking outside mindfully, getting your body moving, experiencing a change of scenery, focusing on a new backdrop, concentrating on the increase in energy that you feel, will restore balance to your system and provide an effective pause for your digitally overloaded mind.

## 5 Unitask

What we do matters. How we do it matters more. Given our extreme multitasking tendencies, working over several platforms and various screens at the same time, many of us are actually finding that we are losing our ability to concentrate on one thing at a time. When we do one thing at a time it can feel slow and unproductive. In fact, concentrating fully and in a focused way on one task increases productivity. Research shows that the effects of multitasking result in us having trouble focusing and filtering information, and lead to increased stress levels. It has further been shown that even when we have finished our multitasking activity, the residual effects of our fractured thinking and lack of focus stays with us for an extended period.

## 6 It's intentional

Establish your intention for the day. By starting the day with a single prioritized focus, this will encourage concentration and create a manageable guide for your day's activities. Ask yourself what you would like to achieve during this day and what quality you want to embody while you are moving towards your intention. We are generally so focused on trying to do as much as we can that we often forget to concentrate on the quality of the individual tasks we are actually undertaking, and the way we are while we are completing them.

## 7 FaceTime: the analogue version

As human beings, we have a primal need to touch and feel and connect in a real way. We have to remember that within us is a fundamental human desire to feel emotion face-to-face and not just through a digital filter. It is important to engage our five senses again and start to touch things other than a digital device. We thrive on facial feedback, as emotionally this gives us a sense of wellbeing through oxytocin being released by the brain. We need to stop finding ways, through digital distraction, to avoid physical contact.

## 8 Mindset reset

Find your own "mindset reset" button. Try to spend at least 5 minutes each day doing nothing; ideally several times a day. First thing in the morning, instead of reaching for your smartphone, checking your emails and scanning through your mental to-do list, try to spend 5 minutes with your eyes shut practising gratitude for all the things in life that you have, that money can't buy. There is no limit to the practice of gratitude; you can fit in a quick 5-minute "mindest reset" anytime, anywhere. Express gratitude, live in possibility, stay curious and be generous.

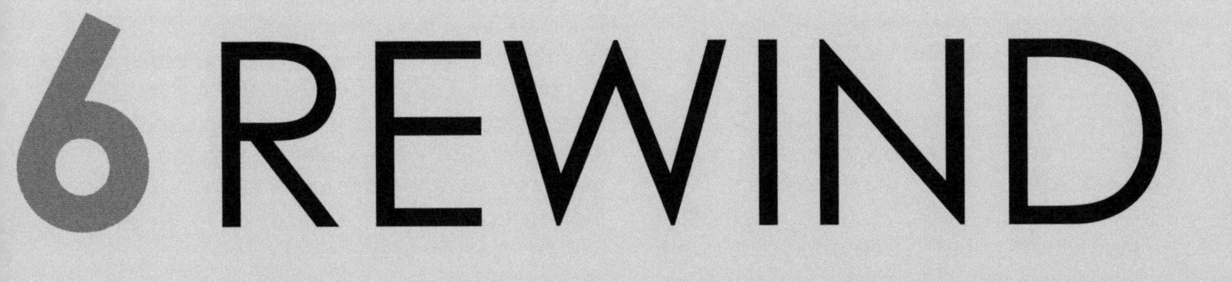

# 6 REWIND

## The digital detox plan

**We've all felt that rising panic when we think we may have forgotten, misplaced or, even worse, lost our smartphones. Therefore the idea of separating from what we have come to view as our "digital lifeline" through choice, by undertaking a digital detox programme, is never going to be an easy process. It is exactly because it is such a challenge that we need to ensure that we find a way to do it.**

However, like any habit or addiction none of us is going to be able to unplug for any length of time because we feel we OUGHT to. We are only going to be able to go tech-free if we WANT to.

We crave being permanently connected and are filled with anxiety when we are not. We struggle under the overwhelming pressure of continuously sharing information. The resultant "malaise" we often feel, due to the extreme levels of digital immersion we experience on a daily basis, leaves us feeling digitally overloaded and yet strangely empty at the same time, because ultimately the sharing of information digitally is not the same as communicating.

We seemed to have reached a point where we willingly trade real communication and human contact for a piece of hardware. A smartphone, does not breathe, or have a pulse or think, but we often give it more attention than we give our partners, children and work colleagues. We give it power by allowing it to monopolize our time, our thoughts and our relationships and, incredibly, we willingly let it survey, track and monitor our every move, without having full knowledge of where that information will finally come to rest.

We have been carried along by a digital wave of change and as a consequence we seem to have lost our ability to experience life in real time. We have lost the ability to find the "spaces in between", filling them instead with digital noise. Managing our digital lives seems itself to have almost become a full-time job. We wear "busyness" as a trophy but in reality our busyness is not productive. None of us are really "too busy". We just need to change the way we prioritize our time and choose living life over digitally editing it on a 2-inch screen.

But perhaps rather than viewing a digital detox as a negative experience to be endured, with the accompanying anticipated sense of loss due to the absence of continuous digital connectivity, we may instead find an unexpected pleasure in the long-term gain we experience through giving ourselves permission to leave our digital dependency behind and reconnect with our own internal "hard drive".

Taking some digital downtime will increase our ability to resist the call of the social network siren, or the magnetism of the message maven and instead focus on the "now". As Ellen DeGeneres said: "All we have is the here and now. That's why procrastination feels SO good. Procrastinate now don't put it off." Humour aside, it is an essential truth that all we have is the "now" and beyond our digitally connected world, the essential truths of life haven't changed:

As Laura Ingalls Wilder says:

**"The real things haven't changed.**
**It is still best to be honest and truthful;**
**To make the most of what we have;**
**To be happy with simple pleasures**
**And have courage when things go wrong."**

Taking the first step is always going to be the hardest part. When you finally decide to "Unplug" for however long feels right for you and your life, the benefits will become apparent almost immediately. During a digital detox, less really is more. You will almost instantly begin to feel less stressed and more relaxed. You will feel less hurried as time slows down and more able to think. Sleep will be less interrupted and more replenishing. Time will be less pressurized and allow for more quality moments.

The following digital detox plans have been formulated to enable you to get a clearer understanding of what you stand to gain by relaxing your grip on the motherboard of our digital life. You will discover that letting go has an upside – it also lets go of you! The plans aim to help you understand that how you manage your technology determines whether it is an enabling tool or a disabling tool. Ultimately it is about adding not taking away, and enhancing your offline life through the reshaping of your online life. The step-by-step process has been designed to help us understand the "how" and the "why" of digital overload in order to make us want to try to find a workable balance between our digital world and our life offline.

A series of detox plans are outlined in this chapter, from a 1-hour tech-free taster (see page 106) for absolute beginners to the equivalent of a 7-day "Unplug and Play" vacation, and everything in between. There are also 10 pre-digital-detox prep steps to prepare for and support your digital detox programme. These practical steps can be integrated into your daily routine and can have a positive impact on restoring the online / offline balance of your life. This chapter has been designed as a series of manageable steps broken down into bite-sized chunks to encourage a re-learning of how to harness the positive and extract the negative from your online life. Each of us is unique, therefore there is no "one size fits all" solution. The idea is to provide a "pick and mix" selection that you can undertake according to your schedule and lifestyle, depending on the time you have available. The detoxes can be done in sequence, undertaken individually, or combined to create a personalized plan that works for you and your life, to help you establish a new practical and workable digital protocol.

Following any digital detox plan will require a commitment and a readiness to deal with change. Change is not easy, but once you've chosen it, if you embrace it you will find that little by little, the rewards will far outweigh the challenges.

# 10 digital detox prep steps

step 1

## Starting Brief

This is an effective first step to get out of our "always on" mode and FOMO (Fear Of Missing Out) mindset and begin to embrace change.

### Preparation time

30 minutes

### Ingredients

- Pen and paper (radical but necessary!)

### Instructions

Find a quiet space, ideally with a table and chair where you can be on your own for 30 minutes. Take a few minutes to focus fully on your life and how it is now. Now imagine for a moment that this was your last day. Take some time to really think about and then write a list of the 10 most important things that you would want to do with your last day.

### Digital detox supplement

Would you want to clear your inbox by answering every one of the 500 pending emails? Or spend time making sure you send every one of the pending emails in your outbox? Would sending that tweet and Instagramming that meal and checking into Facebook be your priority? Would spending an hour surfing for the latest model of something you already have make you happy?

Or would you give anything to actually be able to touch, hug and talk face-to-face with a loved one? Or go for a walk surrounded by nature where you can feel, breathe and hear the natural wonders of the world? Or actually go and visit the people and places you love, in person? We don't have to wait till it's too late. Choose life now.

step 2

## On Point

Identifying the key points and personal goals that you want to achieve during your digital detox are pivotal to its success.

**Preparation time**
30 minutes (one time only)

**Ingredients**
- Notebook and pen (getting back to basics!)

**Instructions**
Focus fully on your life and think about what you would like to achieve during your digital detox and write down your goals in your notebook. Establishing a set of personal goals that are achievable will encourage you and provide you with an incentive to continue with the digital detox plan until you have reached these goals. They could be larger life goals such as spending your newly available "unplugged" quality time with friends and family or starting to pursue a long-held passion, or they could be more practical such as taking your smartphone out of the bedroom at night, or deciding to only check your email once a day at the weekend, or a combination of both.

**Digital detox supplement**
Committing your goals to paper will give your digital detox a focus and will provide an effective guide to keep you on track to achieving your detox goals. Once written down you can carry your notebook with you and refer to your list of goals on a "need to" basis.

step 3

## Write It Out

Keeping a daily detox diary will enable you to monitor your progress and provide you with a valuable record of your digital detoxing journey.

**Preparation time**

20 minutes (every evening)

**Ingredients**

- Diary and pen (rediscovering the joy of handwriting)

**Instructions**

Writing a daily diary entry of your detox experience, at a regular time every day, ideally every evening or at the end of the day before you go to bed, in a quiet space, will help you to focus on any issues that came up during that day's digital detox. Be honest with yourself and include any obstacles that you come across, whether self-generated or due to external circumstances, so that they can be highlighted, reviewed and addressed by you in preparation for the next day. This will help to identify the positive benefits of your digital detox experience as well as any negatives that you may experience during the digital detoxing process.

**Digital detox supplement**

Putting pen to paper to write up your daily diary entry, apart from being a much more connected and considered exercise, will enable you to chart your progress and provide you with invaluable feedback on your digital detox journey. This written record will also serve as a detailed road map for you, for any digital detoxes that you may decide to undertake in the future. The diary will also enable you to deal with any issues that arise, such as not being able to resist checking your smartphone or sending a secret email etc., which you may have encountered during your first digital detox.

step 4

## Wake Up Call

One of the simplest and most important first steps to undertaking a digital detox is to buy an alarm clock.

**Preparation time**
1–2 hours (one time only)

**Ingredients**

- An alarm clock (that is not your mobile/cell or smartphone)

**Instructions**
Lose the phone and sleep alone! Or if you are not sleeping alone at least make sure it is not with a smartphone! If your phone is your alarm clock and you sleep with it next to you, you are also by default extending an open invitation to the entire world to come into your bedroom. We like sleeping with our phones, it gives us a sense of being connected, and of comfort. It also stops self-contemplation, relaxation and creative thought. Checking our smartphones before going to sleep, impacts negatively on our ability to sleep and the quality of that sleep. When we are stressed or preoccupied the sympathetic nervous system responds by making our blood pressure rise, the heart beat faster, and muscles tighten. Looking at a bright screen, however small, just before going to sleep makes the body release approximately 22% less melatonin, which is the hormone that triggers sleep, and will guarantee a night of interrupted sleep.

**Digital detox supplement**
Ideally digital devices should be switched off overnight or put on silent mode and plugged in to charge in another room. This prevents your smartphone from being the last thing you reach for before going to sleep and the first thing you reach for when you wake up. Remember we survive 7 or 8 hours nightly without checking our phones (hopefully!), so an extra hour should not make a difference. In the morning instead of opening your eyes and squinting at a backlit screen, why not try opening the curtains and letting natural light in to balance your body's circadian rhythms.

step 5

## My Space

Make a "go to" space that can become your special sanctuary.

**Preparation time**
30 minutes

**Ingredients**

- Pen and paper

**Instructions**
An important step to reconnecting with yourself during your digital detox is putting yourself back on your "to be" list. Creating both a physical and mental space for you just to be able to "be", where you can pause and have time to think, unwind or meditate, will go a long way towards helping you to reconnect with yourself, friends and family and the natural world around you. Take some time to find a quiet corner and make yourself a sanctuary space where you spend some scheduled uninterrupted time. It can be as simple as a comfortable chair where you can sit by a window for a while and watch the light change, or read a book or gently meditate for a while.

**Digital detox supplement**
Once you have created your "My Space" make sure that you schedule in a time every day to spend an hour there, unplugged. Use that time as a positive step to disconnecting with the world and reconnecting with yourself. Give yourself time to be with yourself. Practise some mindfulness (see pages 134–37) by listening to a favourite piece of music, fully focusing on every note; spend some time listening to the rise and fall of your breath; practise some gentle yoga or just rest your mind and body for a while. Taking some unplugged "me time" even for an hour of rest can actually make you more alert and productive, reduce stress levels and encourage cell renewal. Even taking a 30-minute rest or nap just three times a week, according to a Greek study, has been shown to cut the risk of a heart attack by 37%, and a study conducted by NASA found that a nap of 26 minutes could improve subsequent work performance by as much as 38%.

step 6

## Back in the Box

This is a practical and effective way to keep phones off the table at breakfast and at other meal times, providing the opportunity to concentrate fully "on the moment" without distraction and start the day in a positive and focused way.

**Preparation time**
30 minutes the night before (one time only)

**Ingredients**

- A container with a lid
- Willpower
- Discipline

**Instructions**
Before going to bed find a box, basket or container with a lid and put it in the room where you have your breakfast or other meals. Find a place as far away from where you eat as possible, ideally near the entrance to the room and place the box there. This will become your new smartphone dock for breakfast and all other meal times at home. Before sitting down at the table, everyone should put their mobile devices on silent mode and place them into the chosen container and close the lid. Smartphones can then be retrieved from the container after the meal is finished on the way out of the dining area, ideally to be switched on once you have left the room.

**Digital detox supplement**
Find a container you really like. Try implementing this routine for 21 consecutive days at all meal times. By the end of the period, doing this at every meal will feel more normal than texting your way through a conversation. Discover the benefits of being offline and enjoying your food, being able to have "you" time and connect with friends and loved ones through having and being given undivided attention.

step 7

## Frame It

Capture the moment and "frame it" in tech-free mode.

**Preparation time**
30 minutes

**Ingredients**

- Memory (without a card)

**Instructions**
Instead of reaching for your smartphone every time you see a special moment or an image that you would like to capture, try to capture and remember it, unplugged. Just immerse yourself in the experience by holding the moment and savouring the way it makes you feel by "framing it" in your mind and storing it in your internal memory card. To re-enforce the message you can even create a frame shape using your fingers, and take the equivalent of an imaginary photo which will enable you to engage with the moment fully and to remember it more accurately in the future.

**Digital detox supplement**
This tech-free approach to living in the moment and capturing special experiences by being fully present and experiencing them as a 360-degree multi-sensory experience enables you to really connect and "live" the experience, instead of focusing purely on the visual elements, capturing them on a smartphone, filtering them and posting them on social media. This, while it can provide a fleeting moment of entertainment to your "followers", also means that you have lost the opportunity to fully immerse yourself in the experience by focusing all your attention on trying to "freeze" the moment for others. Sometimes just being alone with an amazing "moment" has the potential to be such a meaningful experience that it does not benefit from being immortalized for posterity.

step 8

## Starting Brief

Establishing tech-free places within the home.

**Preparation time**

Half a day or as long as it takes to relocate installed media such as TVs

**Ingredients**

- Persuasive argument as to the long-term benefits of taking this action
- Electrician or electrical tool kit

**Instructions**

In preparation for cutting back on your digital dependency it is important to establish tech-free zones at home. Try making parts of your home, which tend to be the hubs where people gather, such as the living room and the kitchen, unplugged, screen-free zones. The living room in most homes is ususally dominated by a TV screen and is generally used as an "always on" background to "second screening" by family members using their digital devices while intermittently watching TV. Finding another place for the TV, perhaps in another room, den or spare bedroom, would shift the focus away from a media-dominated environment where family members gather to be "together alone" on their digital devices, and provide a tech-free environment for them to gather and communicate with each other in "unplugged" mode. If complete relocation of the main TV encounters too much resistance, a schedule of agreed digital downtimes and tech-free zones can be put in place as an alternative.

**Digital detox supplement**

Other areas in the house such as bathrooms (do we really need our digital devices while washing?), bedrooms and outdoor spaces, and times spent in the car or on public transport could also be used as triggers for periodic digital down times encouraging interaction and communication with those around you.

step 9

## It's a Turn Off

Turning off push notifications on your digital devices.

**Preparation time**

1 hour (once only)

**Ingredients**

- Willingness to let go of
- Willpower

**Instructions**

Receiving continuous alerts on your digital device is a constant distraction and feeds our FOMO (fear of missing out). It also mentally disables our ability
to focus on one thing at a time without being distracted. The relentless stream of notifications from social media, email and messaging makes it almost impossible for us to put down our smartphones let alone separate ourselves from them. In preparation for your detox, go into your settings on your digital device and disable alerts from all your apps, email and messaging. This will enable you, in the countdown to starting your detox, to start to separate and distance yourself from the constant pull of your digital device and regain control of the amount of attention you are willing to give to it, by checking emails, messages and social media notifications in your own timeframe.

**Digital detox supplement**

Stepping back from a fully immersive relationship with social media in particular will enable you to review your social media habits and gain a new perspective. This will also help you to establish whether the time you generally spend on social media platforms is productive and actually brings positive benefits to your life, or whether there is a need to re-evaluate your relationship and dependence on "being digitally needed" via constant social media and other messaging notifications.

step 10

## Back to Basics

Using basic old-school (offline) methods to prepare for your digital detox.

**Preparation time**

2 hours (once only)

**Ingredients**

- Forward planning
- Patience
- Ingenuity

**Instructions**

Preparing for your unplugged time during your digital detox requires forward planning. Make a list of the daily activities, planned excursions and programme of events that you will be undertaking during your digital detox and go through them individually to evaluate all the things you would normally use your digital devices for, such as checking a location with your GPS, checking the weather, getting a route map, your diary, banking, online recipes and your contacts lists, and make printed copies of them in advance. If you are planning excursions, get a compass, borrow a guide book from your local library or pull out that old box of maps and travel guides. Find your old camera and update the playlists on your iPod. Print related articles that will encourage and support your planned activities. Find old recipe books for some family baking time. Write down the key locations and contact information for where you are planning to go. Make arrangements to meet up in advance, agreeing the location, and make sure you are on time (the way it was always done in the pre-mobile age)! Apart from helping your digital detox run smoothly, time spent preparing the information to support your detox will also serve to engage you and get you excited.

**Digital detox supplement**

Thinking about the practical side of your digital detox will enable you be fully prepared before you press that off button and officially unplug, eliminating the risk of being tempted to plug back in before you have completed your digital detox programme.

## The essential digital detox plan

1-hour digital detox

### The Tech-Free Taster

**Preparation**

- Plan to take 1 hour totally "unplugged"
- Put your phone onto silent mode, put it in a drawer and leave it behind!

**Method**

**Weekdays**

Choose a time to go tech-free for 60 minutes when you have planned to do something else so that it will not feel so challenging.

A morning run or gym session provides the perfect opportunity for you to place your smartphone on lockdown, leave it behind and revel in a tech-free hour where you can just concentrate on the workout. Your daily commute to and from work is another perfect time to be in the "moment" and focus on your journey rather than digitally multitasking by talking (hands-free) and driving, texting and walking or reading an eBook on the train. Being mindful and present on your journey gives you time to think, reflect on your day and arrive relaxed and refreshed.

Lunch times are also the ideal time to go offline. Put your digital device in a drawer and let it go to voicemail. Amazingly, the world will survive without being able to reach you for an hour and you might just find that going for a walk, eating without looking at a screen, really tasting your food, and talking to a colleague while giving them your full attention, can feel surprisingly life-enhancing and fulfilling. You may enjoy it so much that you could be tempted to try it again the next day, and the next!

## 1-hour digital detox

### Weekends together

As weekends tend to be more centred around the home environment, deciding to take a daily 1-hour break from technology can become something that can be scheduled into plans for the weekend. Unplugging with your partner, as a family or with a group of friends for an hour can be as simple as everyone "parking" their digital devices in the "special box" (see digital detox prep step 6) at the entrance to the kitchen, and cooking a meal or baking together, or writing a play or performing some music using instruments instead of an app, or even playing a board game (seems an archaic practice but never disappoints)! Finding new activities to make "unplugging" together a creative time can become something to look forward to.

### Alone

Setting yourself a personal tech-free hour on each day of the weekend will give you permission to do the offline things you enjoy, the freedom of fully immersing yourself in whatever "unplugged" activities you choose, whether it is taking a long bath, going for a walk and really appreciating your surroundings, cooking something from scratch using ingredients you have specially selected, reading a book with actual pages that you turn instead of swipe, or just sitting in your favourite chair and listening to a piece of music, will go a long way to making you feel liberated and restored. You may enjoy your digital downtime hour so much you'll want to extend it!

### Benefits

By exploring the opportunities for new, authentic experiences that this time can give you, even for an hour initially, you will gradually start viewing your downtime as a positive part of your life that encourages you to reconnect with yourself and those around you. You may even find that you discover an unexpected pleasure in searching out activities to undertake while you are tech-free and start considering life from a different perspective by viewing some of your normal daily activities as prime unplugging opportunities.

4-hour digital detox

## The "Time Out" Take Out 1

**Method**

**Morning**

Mornings are the perfect time to stay "unplugged" (assuming you have followed step 4 of the digital detox prep step and have removed your digital devices from the bedroom and have bought yourself an alarm clock) and now wake up listening to the sound of birds tweeting, not reading them on your smartphone. Your bedroom should be one of your sanctuary spaces and is the perfect environment to start your tech-free morning.

One of the keys to getting the most out of your digital time out, particularly in the morning, is to get up early. Try and get up an hour earlier than you would normally. Once you have got over the "shock of the new" you will find that quiet, private, silent, distraction-free, uninterrupted time that is all yours is something to savour and appreciate. Early mornings are so peaceful, it is the ideal time to meditate or write down some thoughts (using pen and paper), have a workout, go for a walk or run, or just have a long soak in a bath. Ideally a morning digital detox is not designed to be a time to catch up with work. This is space to be able focus on your inner life. During this time the online, digitally connected "you" has to be parked to make way for the "unplugged" version of yourself to be able to breathe and find its own voice.

**Benefits**

Getting up early and welcoming the day in your own way will give you the time to factor in the eating of breakfast as a part of your morning "unplugged" ritual. Eating breakfast instead of a hurried "grab a coffee and go", can become a mindful experience to savour and enjoy and form part of a gentle ritual to transition you into your day. You can use your "unplugged" morning activities to establish your own morning ritual and, if you like it and decide to repeat it for 21 days, you will be able to replace your previously digitally overloaded mornings with your new personalized version. Note to yourself: Make a rule not to switch on your digital devices till you have completed your morning digital detox and have left the house.

4-hour digital detox

## The "Time Out" Take Out 2

**Method**

**Afternoon**

Afternoons are generally punctuated by a dip in our energy levels, therefore they are the ideal time to unplug. Rather than trying to override the fatigue and lack of productivity we generally experience mid-afternoon we should allow our body's natural biorhythms to guide us in managing our energy levels. Booking an afternoon out during the week to go "tech-free" will normally need to be planned in advance, while scheduling an unplugged afternoon over the weekend will obviously need less forward planning. In many environments "power naps" of 20 minutes or so are encouraged to restore energy levels and this concept can be factored into your unplugged afternoon.

If in a work environment, start by sending out an email notification to colleagues and clients with an auto reply message saying that you will be "away this afternoon and will be picking up emails in the morning". Then switch your digital device onto airplane mode or turn it off altogether and place your smartphone either in the specially designated container (see digital prep step 6) if you are at home, or lock it in a drawer or locker if at work. Power down your desktop computer, laptop and other digital devices and put them out of sight or leave them behind. Step outside and go on an inwards journey. Go for a walk and really look at the natural wonders around you and if in an urban environment notice how the light falls on a building, or inhale the smell of flowers as you walk past a florist, or breathe in a waft of perfume as someone walks past; engage your senses. We are told not "to sweat the small stuff" but we do have to notice it. Now that we are "looking up" and around us instead of down at a backlit screen, we can notice and appreciate the world – the life etched in people's faces, the fleeting moments that make us happy, the sounds that surround us, and really appreciate being here.

# 4-hour digital detox

Doing something as simple as going for a walk can be such an enriching experience. It also gives us a chance to think and allows space for us to connect with our feelings. If spending an afternoon unplugged together with a partner, friends or family, the simple activity of sharing a coffee or tea or a meal and really talking and listening and fully focusing on each other, can be so rewarding. An unplugged afternoon can be as full or empty as you want. You can arrange to be occupied for the entire 4 hours by organizing activities to avoid spending any "unplugged" time with yourself if that feels more comfortable at the early stages of your "unplugging" journey. Alternatively you can use those "unplugged" afternoon hours to power down and mediate, contemplate and reconnect with the analogue, quieter version of yourself.

**Benefits**

By creating a personalized "unplugged" ritual even for a morning or afternoon, you are actually restoring structure into your life that previously would have been consumed and eroded by technology. Your tech-free routine can become a direct line to a more mindful and contemplative "you" where you can step away from your daily routine and connect more deeply with the internal narrative of your life.

24-hour digital detox

## The 1-Day Wonder

**Preparation**

- A popular way to start a 24-hour digital detox is to choose a day that has some meaning for you like an anniversary, birthday, or a landmark day, or if you prefer you can just select a day that works for you on a practical rather than a symbolic level. Try to find a day of the week when taking an "unplugged" day out will not be too complicated. Obviously weekends are often the preferred choice for undertaking a 1-day digital detox.

- In preparation for your "digital absence" always let people know of your imminent "away day" from technology and set up an "away" message on your email account. If you choose a weekend for your 24-hour digital detox, arranging to respond on Monday is completely acceptable. Let friends and family know of your planned time unplugged and make arrangements to meet in advance of starting your detox.

- Ideally the night before you start your 24 hours unplugged, switch off all your computers and digital devices and store them out of sight. Have a look at your list of personal goals (you will have written these down during digital detox prep step 2) and decide what you would like to focus your attention on. Double-check that you have everything that you need before you unplug. Print instructions, meeting points, maps, guides and whatever other tech-free items that you will need to make your unplugged day run smoothly.

- As you have a whole day at your disposal, your day may feel unnaturally long without the continuous distraction of your "always on" life. Make plans, ideally in advance, of what you are going to do during your digital detox, so that you have a day of activities to look forward to. This is the perfect opportunity to get physical and reconnect with your body because ultimately we are all more than our thumbs!

# 24-hour digital detox

**Method**

There are plenty of physical activities that will help you to reconnect with your body. You can start by replacing emails with exercise, push notifications with push-ups, and instant messages with massages. Walk, talk, run, swim, play your favourite sport, get active, get physically tired and relax.

These 24 hours are determined by you and during this digital detox day you control your time, your attention, your thinking space and your focus. Wherever you decide to go away or stay at home make sure you are fully present in everything that you do. The 24 hours of digital down time can also be a gentle, meditative and nurturing time, depending on what your priorities and needs are at the time.

Make sure to factor in some down time and some "alone" time so that you give yourself an opportunity to slow down, give your brain a chance to recover from digital overload and provide you with an opportunity to be fully present wherever you are. If you plan to go on a trip for the day then try and include some "low key" moments amongst the physical activities.

At the end of the day take some time to unwind. Create space to meditate, have a long bath to soak those aching muscles and read a book before going to bed. Being unplugged 24 hours can end up feeling as if you have had a "life in a day". You will have experienced life in 360 degrees, you will have worked your body, relaxed your mind, reconnected with loved ones and given your inner life a chance to breathe.

**Benefits**

After undergoing a 24-hour digital detox you may have many mixed emotions. The relief and freedom you feel through periodically cutting the digital umbilical chord may ultimately turn out to be more addictive than your need to be connected to it. There may also be a euphoric feeling that you've tasted and been reminded of what it is like to really live and be present for a day, albeit tinged with the sadness of realizing what you have been missing. When you do finally log back on you will be surprised at how little you missed.

48-hour digital detox

## The "Web-free" Weekender

### Preparation

- Weekends are often the best time to unplug, given that your time is much more your own, and therefore planning to have an "unplugged" weekend is less complicated and can be arranged in advance. Whether you decide to stay at home or go somewhere for the weekend, it is helpful to prepare a list of things you would like to accomplish and activities that you would like to do during your tech-free time (see digital detox prep step 2). These should also include some quiet time to be on your own, and quality time to spend with friends and family. Given that there are two days, you can alter the rhythm and intensity of the days so that at the end of the weekend you will have had both socially active times and more introspective moments that allow you to have time with yourself.

- In preparation for your weekender, switch off all your computers and digital devices and store them out of sight. Double-check that you have everything that you need before you unplug so there will be no excuse to plug back in. Print instructions, meeting points, maps, guides and whatever other items that you will need to make your weekend run smoothly.

- In advance of your "digital absence" always let people know of your imminent weekend away from technology and set up an "away" message on your email account to let people know that you will be logged off and away for 48 hours and will respond to their emails and messages on Monday. Let friends and family know so they are not alarmed when they cannot reach you and make arrangements to meet in advance of starting you detox. For emergencies let selected friends and family have a contact number where they can reach you.

## 48-hour digital detox

**Method**

Plan some family time doing the sort of things you did as a child: play games, draw pictures, go to a park, climb trees, play ball; you can even collect mementos and souvenirs like food wrappers or bottle lids and make a scrapbook of your day together when you get home. Buying ingredients locally and preparing a meal together and eating it mindfully is a great way to connect with friends and family and will give you an opportunity to enjoy having meaningful conversation with those around you. Spend some time in the garden together or go for a walk in a park and really look at the trees and allow yourself to be fully immersed in nature's details. If you do decide to take a trip it doesn't have to be somewhere far away; it can just be somewhere local that you've never been to before and have always wanted to visit, or somewhere you know but haven't really taken much notice of as you were fully focused on your smartphone.

Remember to be mindful, don't miss the "moments". Life is full of special fleeting moments that together make a life. Don't miss them in your quest to always paint the bigger picture or dismiss them as insignificant unless they are Instagram worthy!

The purpose of going tech-free for a weekend is to be in "the moment" to be really present and start noticing all the things that you have been missing while immersed in your smartphone.

**Benefits**

You will start to find that disconnecting for a weekend from technology is a nurturing experience. Enjoy the down shift in the pace of your day and take pleasure in no longer having to be on "read alert" as an instant-response squad to incoming messages; focus on just dealing with the now. View your weekend as a gift to yourself, a reward that delivers you back to yourself.

7-day digital detox

## The "Unplug & Play" Holiday

**Preparation**

- Unplugging for a week takes planning. Often the best time to take a 7-day digital detox is to coincide it with a scheduled holiday or break. Whether you decide to stay at home or go somewhere for the week, it is helpful to make a note of the things you will need to prepare for your tech-free week and write a list of the things you would like to accomplish while you are unplugged (see digital detox prep step 2).

- Whether you decide to stay at home or use the time to take a holiday, alone or together with friends and family, a 7-day digital detox will give you the opportunity to step outside the digital techno bubble that we are surrounded by and give you a chance to really connect with yourself and those around you. Being unplugged for this length of time will enable you to prioritize and rediscover the things that are important when you disconnect.

- In preparation for your week, switch off all your computers and digital devices and store them out of sight. Make sure that you really have everything that you need before you unplug so there will be no excuse to plug back in during your 7-day tech-free time away. Print instructions, meeting points, maps, guides and whatever other tech-free items that you will need to make your "holiday" run smoothly.

- In advance of your time away, always let people know of your imminent digital downtime. Brief co-workers and ensure that they really understand that you will not be available. Set up an "away" message on your email account to let people know that you will be logged off and away for 7 days. Let friends and family know of your planned time so they are not alarmed when they cannot reach you and make arrangements to meet in advance of starting your detox. For emergencies, let selected friends and family have a contact number where they can reach you.

- Resend the notification message the night before before you leave or start your digital detox just to remind everyone that you have gone "dark" and will be off the grid for 7 days and give them the return date when you will be reconnecting and logging back on.

# 7-day digital detox

- Establish a list of priorities for your 7-day digital detox so that you will be able to allocate enough time to work through each of the items on your list, as you will feel as if you have so much more time than usual because your time will not be occupied by mindless engagement with your digital devices.

**Method**

Try to include a programme of activities for both the mind and the body, to be undertaken both alone and together with family and friends, in order to balance and enhance your physical, mental and emotional wellbeing.

Plan to try and engage all your senses by scheduling some form of physical activity every day that activates the body, such as walking, running, swimming, yoga or playing a sport.

Allow some quiet time every day for some meditation, contemplation, reading a book or writing some of your thoughts down. Unplugging will give you an opportunity to think beyond the present moment.

Breathe. We all breathe without consciously being aware of our breath. Give yourself some time to concentrate on the breath and practise some deep Balancing Breath techniques (see page 79). This will help you to feel grounded and will increase your focus by calming and balancing the nervous system. It can either be done separately or as part of your daily meditation practice.

Make sure you factor some "slow" moments where you live more mindfully and notice the finer details of life, the people you are with and your surroundings.

Make memories that can't be posted online but will be remembered in your own internal memory card. Schedule some group or family activities where you are all engaged in an activity or all go on a day trip and can fully interact with each other. Rediscover the joy of simple pleasures and have some fun!

Spend meal times together really communicating and enjoying each other's company, and giving those you are with your full attention.

## 7-day digital detox

If away for the week, take pleasure in visiting a local market or independent craft stores and search out some handmade finds with provenance and a back story that you could never have found online.

Take some photos of where you are and the people you are with on a digital camera that doesn't have an inbuilt selection of filters and editing tools. Enjoy taking pictures of each other (rather than of yourself) and get them printed on photo paper when you return home and send out copies to everyone who shared your time away. These prints could even become the focus of your post-unplugging reunion, when you could gather to make the photo album (traditional style with paper pages).

**Benefits**

When we step back and take a break from our over-consumption of digital media for a while it throws our previously over-connected, noise-filled life into sharp relief. We spend so much time overrun by excessive amounts of digital information that we have no time to manage our thoughts, always feeling like we are running but never arriving at a point where our lives are under control.

As with all habits your digital dependence is a habit that develops through repeated use. So, much in the same way, having spent 7 days getting used to being separated from your digital devices can become a habit also. To make lasting change takes commitment, self-discipline and repetition.

After a week offline you will have experienced a new sense of freedom, a new gentler time frame that allows for pausing and breathing and feeling again. Gaining a new perspective through having lived an alternative, more authentic and connected version of your life for a week should leave you feeling renewed, revitalized, less willing to multitask and multiscreen. Perhaps it will encourage you to make a commitment to restricting the digital white noise in your life and continue to step back periodically and disconnect from digital media on a regular basis to allow space in your life for real and meaningful connection with yourself and genuine interaction with the people and the world around you.

# Conclusion
# RESET

**In the pre-digital era we let life happen without having the ability to edit it as we went along. We now live in a world where we have a dropdown menu of life-editing choices, offering us an overwhelming array of options for controlling, sharing and filtering our enhanced online lives. However, that throws into sharp relief the aspects of our lives that we can't control – our daily, challenging and imperfect analogue world, the one that we physically live in. If someone could promise you a perfect, ironed, crease-free, edited life instead of the one you are living, would you take it? A life where the volume was always set at five, where the experiences are all under your control and the way people perceive you is edited to ensure you present yourself perfectly at all times? Would you trade your real life for an iLife? By continuing to prioritize our digital world over our offline world, that is essentially what we are doing.**

Maybe our digital world is a reality where we are living but not present in our lives. We are literally missing in action from our analogue lives. We are going through the motions of being physically present but mentally absent, by being fully invested in our smartphones. We are walking through life with our eyes blinkered, with our senses on "mute", our daily lives being directed by a piece of hardware.

We spend our time online creating continuous digital distractions and diversions. It feels far more comfortable to be busy with our online activities than to feel vulnerable because we are forced to be alone with ourselves. We are beginning to realize that the good parts of life cannot be captured and held on "live pause" because everything is transient. The nature of life is that it is uncertain, yet we try to control it digitally. It is precisely because we understand the impermanence of life that we have a desire to hang on so tightly to all that seems stable and try to capture and immortalize it online, upgrading the transient to the permanent.

We spend much of our time online digitally joining the dots of our lives to avoid living the spaces in-between, the real, raw, sometimes lonely places that we replace with continuous connectivity. We want to control the messy bits by rewriting the challenging moments and enhancing the good ones. We want to share our best profile photo rather than our best selves, turn up the volume and increase the saturation of the way our daily lives are perceived. That is a full-time job that leaves little time to experience life while we are living it.

We have to learn to respect the process of letting life unfold in its own time and in its own way. It is precisely the randomness and unpredictability of life that makes it the worthwhile adventure that it is. Ultimately the wonder of being alive is found in the unexpected, the "not knowing". Life will not mould itself to our intended roadmap. Instead it is through life's off road experiences that we grow and evolve.

Our lives are buckling under the weight of digital information overload. We now skim through information rather than taking the time to read it. We scrape the surface of content by grazing on information presented in sound and vision bites. We answer using reductive "text speak". We are living our lives in headlines, without the rest of the story. We define ourselves by the online profile and the edited content that we share. Are we becoming what we post? Does a reduction in the way we define ourselves ultimately diminish us as human beings?

We view instant digital gratification as the gold standard of being. We search out instant solutions and immediate responses. We think faster is better and slower unproductive. We are used to having everything we want in the palm of our hand but is it everything we need? We treat digital devices as a lifeline. But does our continuous digital connectivity really support us and our lives or is the time, energy and attention it burns becoming the fuel that drives it? When did we reach the point where we willingly trade human contact for a piece of hardware?

When did who we are become not enough? We accept that technology has extended and amplified our experience of the world by adding an extraordinary and empowering layer to our lives. However it is our relationship to it that will determine whether it is a positive facilitator or a reductive debilitator.

Our relationship to our online world is determined by us. Our digital devices can only have as much control and power as we give them. We have a choice. We can learn digital manners and adopt a new digital protocol so that we can restore balance to our lives. This will enable the living, breathing and feeling elements that make us human to find their way back to our "to be" list, so that we can replace our digitally reductive "soundbite" of a life with a life where we can be everything that we are and are capable of being, ultimately becoming more by connecting less.

Unplugging is not so much a disconnection as a fine-tuning of our inner search engine. Practising moderation by finding a workable balance between our digital connectivity and our real-life connections is the key to establishing a new digital protocol where we can be fully present in our lives while using our digital devices as the tools they were designed to be. We accept that we live in a digitally dominated world that we have allowed to impact on every aspect of our lives, but although we are living digitally curated lives, ultimately we can't Photoshop our personality, or edit our souls.

We have to establish a new modus operandi of how to live with this new mind-altering, life-changing addition to our lives. Ultimately I believe that we know all the answers to the questions we want to ask, because they have always been there inside of us. We just need to create enough space in our lives and pause for long enough to be able to hear the answers.

We are living, breathing and extraordinary human beings with five senses, so let's find new ways to reconnect with ourselves by making space to be able to be fully present. By embracing rather than erasing solitude, we will be able to live a more conscious and mindful life where we can allow the "now", and through presence gift ourselves back to ourselves and to those who have a special place in our lives. We are here, now. Let's live the moment and rediscover the joy of mindful living in a digital age by finding a new way to reconnect with the poetry of life.

# THE TOP DIGITAL DETOX RETREATS

## KAMALAYA, KOH SAMUI, THAILAND

One of the most exquisite, peaceful and spiritual places I had the privilege to visit during my research for this book is Kamalaya. This breathtaking retreat is nestled in a lush green tropical pocket on the beach in Koh Samui, a jewel of an island off the southern coast of Thailand.

This multi-award-winning Wellness Sanctuary and Holistic Spa is an expression of founders John and Karina Stewart's life experiences and their desire to serve and inspire others. Founded by them in 2005, Kamalaya offers a holistic wellness experience that integrates healing therapies that draw from both the East and West. Kamalaya's beautiful natural environment, inspired healthy cuisine, holistic fitness practices and customized wellness programmes make undertaking a digital detox here a unique experience.

Kamalaya feels as if it evolved organically from a vision rather than a blueprint, and was born from a profound passion to create a sanctuary space. Kamalaya's essence is about harmonizing and healing. It is founded on spiritual and holistic principles and offers a rare form of retreat that is increasingly hard to find. A place where you can really unplug.

Co-founder Karina Stewart, a renowned doctor of Chinese medicine, noticed more and more guests coming to Kamalaya with adrenal burnout and sleep issues. She views both of these conditions as health imbalances that are often related to a fast-paced lifestyle in which people's reliance on digital technologies to keep in touch and stay connected 24/7 is a significant contributing factor, which can lead to disruptive sleep patterns and more serious long-term health issues.

At Kamalaya guests are encouraged to abandon the habit of regularly using their digital devices, so the use of mobiles, laptops, iPads, etc., are prohibited in all public areas of tranquillity such as the Wellness Sanctuary and The Alchemy Lounge. "Taking a break from all things digital, be it iPhones, laptops, any digital devices, is essential if you want a deeper digital detox. Undertaking a physical and energetic detox is necessary in order to eliminate the toxins absorbed from close proximity and prolonged use of these devices," says Karina.

Kamalaya makes a point of not offering a specifically named "digital detox" treatment within their wellness programmes, as they believe that all their guests will naturally undergo a digital detox as they are encouraged to reconnect with nature and let go of their digital addiction, whatever wellness programme they choose. All the comprehensive Detox Programmes bring together medical science and holistic therapies.

While I was there to undertake my own digital detox, I experienced their 5-day Asian Bliss programme – a holistic and healing plan that included an initial wellness consultation, a Bio Impedance analysis and a fusion of Eastern therapies that included meditation, Ayurvedic massage, traditional Thai massage, Chi Nei Tsang, reiki and yoga. With each treatment I felt the importance of things like tackling my inbox, getting my messages and sending emails recede and their relevance begin to diminish.

Beyond the actual treatments, which are so nurturing and replenishing, it is the spiritual, life-enhancing and gentle atmosphere at Kamalaya that I found goes such a long way to encourage the restoration of balance and the finding of new meaning and connection with ourselves and our natural world.

Even though I was using a digital device to write this book during my stay, as each day passed I found that I was beginning to resent the whole idea of taking a cold, shiny, hard, metal iPad out of my bedside drawer. I found myself increasingly delaying the time when I would start writing, as I was revelling in the fact that I could just be without anyone or anything making demands on my time or attention. It completely changed my perspective from the inside out and from the outside in. The more I reconnected physically and emotionally with myself and inhaled the exquisite surroundings, absorbing the tranquillity of this harmonious oasis, the more I started viewing my digital device as just that – a device, a tool, an inanimate piece of hardware. I had started to regain perspective, and from my new standpoint the idea that I had given that device so much power seemed so incongruous.

Kamalaya is a peaceful sanctuary, a paradise that is both spiritual and authentic. It is a retreat that encourages us to disconnect from our digital life and embark on a life-changing journey to reconnect with the essence of life and rediscover who we really are.

During my stay I also had the opportunity to meet with and interview Karina Stewart, MA, TCM (Traditional Chinese Medicine) and co-founder of Kamalaya Wellness Sanctuary & Holistic Spa Resort.

**1. Kamalaya has at its core a breathtaking Buddhist monks' cave that was originally used as a spiritual retreat and a place of meditation. Was finding this spiritual cave the catalyst for the creation of Kamalaya?** Everything that my husband John and I have learned and experienced in our lives inspired us to create Kamalaya. We dreamed of creating a place where healing and spirituality could naturally co-exist, using our own life experiences and training to make a positive contribution to the lives of others.

The original vision for Kamalaya began when John and I married and lived in Kathmandu, Nepal, and we were planning to create this place there. But during a visit to Thailand in 2000, John was drawn by the healing nature of Koh Samui and found the site for Kamalaya. Koh Samui has long been favoured by Buddhist monks as a sanctuary for spiritual retreat. Once we found the cave we both knew that this was the place to build our dream.

**2. As the co-founder of Kamalaya you are also a doctor of Chinese medicine. Does your expertise in the field of alternative medicine form part of the unique holistic treatment programme on offer at Kamalaya?** Absolutely. Traditional Chinese Medicine as well as other ancient medical traditions like Ayurveda provide the foundation for Kamalaya's holistic healing concept that acknowledges and incorporates the healing power of nature. Health and healing are understood as the natural result of living a life in balance that includes mind, body and spirit. And reconnection with nature, the elements, one another and, most importantly, oneself are vital in achieving that balance.

**3. What do you think are the long-term consequences of continuing to lose our connection with ourselves and who we really are through our digital over-connection?** The loss of connection to oneself and to loved ones, to our communities and our environment are only some of the consequences of our immersion and overuse of digital technologies. When we lose our connection to ourselves and to our loved ones, we run the risk of losing our essential humanity – our ability to know ourselves, to identify with others, love one another, our compassion and kindness, all the things that make us human. We also risk losing sight of what really is important in life, and need to be made aware of the damage to our emotional and psychological wellbeing that over-connection and over-dependence on digital technologies can cause.

**4. You talk about our "happiness index". What do you think are the key factors that drive our**

**happiness?** One of the ways we encourage people to move more fully into their emotional framework and away from the intellectual paradigm is to create an environment that encourages each of us to become more immersed in the present moment physically and more aware of our senses and nature all around us. We also create a safe and nurturing environment, where guests are able to relax deeply and explore emotions and feelings that may arise, and we provide support through that journey.

**5. Do you think nutrition has a role to play in our journey from digital overload back to wellness?** Nutrition can ameliorate the damage done by digital overload/overuse by replenishing the body with the nutrients necessary to balance the nervous system. In particular, B vitamins, minerals, antioxidants will be very beneficial, especially when supplied by a plant-based diet. Herbal teas and spices as condiments also play an important role in providing added phytonutrients, to rebalance the effects of over-stimulation of the nervous system, and decreasing the use of sugar, a substance considered more addictive than cocaine and as addictive as heroin, especially when combined with caffeine and digital overuse.

Food is a fundamental aspect of Kamalaya's holistic approach to health and wellness. While it can be hard to learn to control our emotions or our mind, it is fairly easy to control what we eat. By opting for a healthy diet on a daily basis we can make a huge difference to our overall health and wellbeing. Some of the oldest Asian healing philosophies, such as Traditional Chinese Medicine and Ayurveda, have always looked to food as the medicine of choice.

**6. At Kamalaya you offer what you describe as "Asian Alchemy", which is a fusion of healing, therapeutic practices and philosophies that "can lead to profound transformational experiences". Is this one of the programmes you would recommend for the treatment of digital overload?** With "Asian Alchemy" we describe the fact that Kamalaya brings different elements together that work in synergy to help guests reach their individual wellness goals in an optimal way. The focus is on how the different treatments, therapies, activities, environmental aspects, cuisine and people come together to create a synergistic wellness experience. Guests wishing to change their "online habits" would be candidates for several of our programmes, including Basic and Comprehensive Detox, Asian Alchemy, as well as Balance & Revitalize Basic and Comprehensive.

**7. Meditation forms a core element of the awakening and healing programmes at Kamalaya. You offer a variety of meditation styles. Is there a particular form of meditation that is especially suitable for people suffering from stress and adrenal burnout, one of the side-effects of a lifestyle that is "on" 24/7?** Our practitioners teach a variety of meditation techniques depending on the individual's experience, requirements and their goals. Practices may be as short as 5 minutes for beginners to longer meditations for those with more experience. We have meditation experts who are trained in various styles suitable for beginners and advanced students alike.

**www.kamalaya.com**

## THE MANDARIN ORIENTAL, BANGKOK, THAILAND

Luxury resort hotels are not usually the first option to come to mind when in search of a digital detox. However, there are exceptions, and the Mandarin Oriental Hotel is one. Known as the "grande dame" of resort hotels in Bangkok, the historic Mandarin Oriental stands majestically on the banks of the Chao Phraya River and was the first hotel to be built in Thailand, officially opening as "The Oriental" in 1879.

The Mandarin Oriental is a hotel of "firsts". The

Oriental Spa was the first spa to open within a hotel property in Bangkok. Since its opening in 1993, this award-winning spa has been widely recognized as a pioneer in the Thai spa industry. With its mission to provide an all-encompassing spa experience that couples traditional 2,000-year-old Thai holistic teachings with contemporary philosophies, it has successfully created a harmonious atmosphere that is both nurturing and restorative.

I first visited "The Oriental" as it was called then, 20 years ago with my grandparents at a time when the "floating market" still could be found gliding along the banks of the Chao Phraya River. So returning to it with such vivid memories held great personal meaning for me. It often feels risky to go back to somewhere that holds such special memories, but the Mandarin Oriental did not disappoint. If anything it felt as gracious, elegant and noble a place as it had done all those years before.

For my personal digital detox programme I was recommended The Life Balance Ayurveda Programme at Oriental Spa, a course of treatments mainly aimed at reducing stress and its impact on the body and mind. All the therapies at the Oriental Spa are based on a holistic approach, combining meditation, massage and the use of natural herbal remedies, an ancient tradition in Thailand. The spa itself is housed in an exquisitely restored 100-year-old teakwood house, with antique Siamese décor, and is a real oasis of peace and tranquillity.

My experience of The Life Balance Ayurveda Programme was exactly what I needed to counter the effects of my digital overload, and for me served as the perfect introduction to rediscovering how to achieve "Life Balance" through a combination of Ayurveda treatments and Yoga sessions. The aim of the programme is to restore, relax and re-energize the self, facilitating a deep detoxification and reconnection with oneself. For me this treatment programme did exactly that and also served as a much needed "pause", enhanced by the sublime surroundings of the historic spa building.

**www.mandarinoriental.com/bangkok/**

## SIX SENSES YAO NOI, THAILAND

Located on a private island and set in the exquisite natural surroundings of the famous limestone pinnacles of Phang Nga, this Six Senses escape is the ultimate expression of barefoot luxury coupled with sustainability, providing the perfect environment in which to disconnect from our digital world.

The Six Senses Yao Noi is the epitome of relaxed Asian style, offering total seclusion and featuring the renowned "Six Senses" spa experience with its signature holistic treatments. The Six Senses Ya Noi provides the perfect combination of a secluded, discreet retreat coupled with a relaxed and unhurried atmosphere, encouraging a total reconnection with the natural rhythm of life. Optimal nutrition to support the personalized wellness programme is provided by the fresh ingredients grown in the resort's garden, which are used to create organic gourmet dining.

Using only naturally and ethically produced spa products, the spa's treatment menu promotes traditional Thai healing practices and incorporates the four elements of earth, water, fire and air, to ensure the senses are balanced. As part of the holistic wellness programme, Six Senses resorts now also provide a Yoga and Yogic Detox treatment programme.

An additional Sleep treatment will also be incorporated into the Six Senses Spa Yogic Programme series, to re-educate the body and encourage healing and restorative sleep without the disruption of digital noise. This reflects their commitment to delivering holistic, yet result-driven therapies, which bring together the expertise and skills of in-house yoga masters with the healing properties of the surrounding environment.

**www.kuoni.co.uk/thailand/small-island-hideaways/hotels/six-senses-yao-noi**

## RANCH AT LIVE OAK, MALIBU, CALIFORNIA, USA

Given that The Ranch at Live Oak feels like the perfect escape, set amidst acres of native grounds, with hiking trails, picturesque mountain and canyon views, and authentic natural splendour, it seems surprising to learn that it is only an hour away from Los Angeles. Nestled in nature with no distractions, the Ranch's mission is to get their guests to subscribe to their "unplug and off the grid" philosophy, ensuring that their stay will be "memorable, satisfying and highly rewarding".

Located 3 miles above the Pacific Ocean in the Santa Monica Mountains, the Ranch is housed in a Spanish hacienda and features The Ranch's great room, their own organic farm and kitchen, and an exercise pavilion. The Ranch is the ideal place to reconnect with nature and offers the perfect setting to focus on personal health and wellness goals. During a week's programme guests cleanse their minds and bodies by spending days hiking in the surrounding mountains, doing yoga and having post-workout massages, and eating organic nutritious food. In addition to their strict no mobile (cell)/smartphone or Wi-Fi policy, sugar, caffeine and even clocks are also banned.

Their strict policy which allows for a complete and total disconnection from the "grid" encourages guests to also "unplug" from their normal daily activities and instead focus their energies towards being present in their natural serene surroundings.

A stay at the Ranch is holistically designed to ensure that guests will have a profound and meaningful experience, which will propel them to continue on their personal path to health and wellness once they return to their daily lives.

**www.theranchmalibu.com**

## BANYAN TREE SPA SANCTUARY, PHUKET, THAILAND

Bordered by the gentle waves of the Andaman Sea, Banyan Tree Spa Sanctuary is literally a sanctuary for the senses, tucked away in a tranquil corner of Laguna Phuket, Thailand. This sublime spa resort focuses on wellness with its programme of Ayurvedic treatments, organic food and daily yoga classes, and seems to be the perfect place to disconnect from our digitally dominated world and reconnect with ourselves. It is the ultimate escape from all things digital and offers a holistic approach to wellbeing set in this beautifully relaxing hideaway.

An oasis of serenity in a secluded setting, this exquisite and unique spa resort focuses on holistic sensory experiences, healthy cuisine and wellness activities for complete rejuvenation and relaxation. It is the epitome of natural luxury.

The award-winning Banyan Tree Spa offers a range of especially formulated treatments and holistic therapies, with a special focus on creating physical, mental and spiritual harmony. Upon arrival a therapist will escort guests to their villas for an introductory foot treatment, and during their stay, guests will also have the opportunity to meet with the Ayurvedic doctor.

Holistic wellbeing at this beautifully relaxing hideaway includes restorative spa therapies within your personal treatment pavilion and nourishing spa breakfasts freshly prepared within your villa. Complimentary daily yoga classes take place in the resort's own Orchid Garden.

As the first luxury oriental spa in Asia, it introduced

an exotic blend of ancient health and beauty practices, which have been passed down from generation to generation.

Banyan Tree Spa Pavilions are re-creations of Royal Thai salas and combine the splendour of their historical heritage with the beauty of exquisitely crafted natural local materials.

This is the perfect place to reconnect with all the senses and have a chance to restore balance to the mind, body and soul in the most exotic and peaceful surroundings.

**www.kuoni.co.uk/thailand/phuket/hotels/banyan-tree-spa-sanctuary**

## CAMP GROUNDED, ANDERSON VALLEY, CALIFORNIA, USA

Digital Detox, an organization founded by Levi Felix, created Camp Grounded as a digital-detox summer camp for adults. Camp Grounded was conceived as a place where people can come to be separated from their devices, start to learn to treat their digital dependency and rethink their attachment to their digital devices.

Over the course of four days "Campers" unplug and embark on a series of organized outdoor activities set amongst the Californian redwood forests for an "off-the-grid weekend of pure unadulterated fun". It is less a spiritual journey and more a "throwback to childhood" weekend, albeit with some mindfulness, yoga and meditation sessions punctuating the physical activity programme.

Participants are encouraged to spend four days undertaking unplugged activities, such as immersing themselves in nature by staying in a camp in the woods, spending time meditating and discussing openly their relationships with technology.

People at Camp Grounded do not use their real names and are not allowed to talk about their professional status. Networking is strictly off limits. Camp Grounded is not only proudly "off the grid", but over the course of a weekend it also aims to encourage a broader re-evaluation of how we live as a society and the value systems we subscribe to. At Camp Grounded the aim is to "create a community where money is worth little, and individuality, self-expression, friendship, freedom and memories are valued most", bridging the gap between disconnecting digitally and reconnecting with ourselves, our natural world and each other.

Since its launch, the demand for Camp Grounded camps has exceeded all expectations and has led to Levi Felix launching his brand of digital detox camps further afield in Bali, Nicaragua and Cambodia via his company Digital Detox.

Less of a camp and more of a retreat, Digital Detox provides, according to Felix, "a unique opportunity to reduce your stress and anxiety as well as your tech dependency". With smaller groups (10 people) and a range of gentler activities such as yoga, meditation, swimming, art and writing workshops, and, most importantly, writing a daily journal of your unplugged activities, these digital detox retreats will explore the benefits of unplugging while providing "take out" techniques that will encourage us to find balance when we return to our digitally-driven lives.

**www.campgrounded.org**

**www.thedigitaldetox.org**

# Useful Apps

**Apps for increasing focus and productivity at work.**

**www.macfreedom.com** Freedom is an Apple app which turns off your internet, for a pre-set period of time, to help restore productivity.

**Pomodoro Timer Lite** Divides up work into 25-minute chunks, with a short break in between, to manage your tasks more efficiently.

**www.Focustimeapp.com** Groups tasks into uninterrupted blocks of time.

**Evernote** A multipurpose app that lets you create to-do lists, write notes and tasks that sync across all your devices.

**www.sanebox.com** Sanebox provides the ultimate inbox control by intelligently analyzing your emails and automatically filtering out spam and unimportant messages to only leave the emails that are important.

**Focus Lock** Blocks all distractions from your phone so you can focus on the tasks at hand.

**Moment** An iOS app that automatically tracks how much you use your phone. You can set daily limits yourself and be notified when you go over.

**NowDoThis** A web app that helps you to create a one-at-a-time to-do list by controlling multi-tasking by letting you input a task list and only showing you one at a time so that you have to complete the first task in order to move on to the next.

**Apps to help you be more present, de-stress and become more mindful.**

**www.headspace.com** Headspace is an app that was co-founded by Andy Puddicombe, a former monk, who has created a set of guided meditations to be practiced for
10 minutes every day. This is the perfect Meditation "to go".

**www.buddhify.com** Buddhify has been hailed as the "urban meditation app" described as "Modern mindfulness for wherever you are".

**www.franticworld.com** Franticworld is an app that provides a selection of free meditations for mindfulness and includes several different versions of meditations.

**www.calm.com** Provides online meditations in "mini sessions" of 2, 5 ,10, 15 and 20 minutes, and the app provides relaxing music timers to customize your meditation practice according to the time you have available and video content of inspirational natural scenes.

**www.chopracentermeditation.com** Oprah Winfrey and Deepak Chopra joined forces to launch a 21-day online meditation experience, called Finding Your Flow. This profound and transformational online path to meditation has reached over 2 million people from around the world.

**www.stillnessproject.com** The Stllness Project is a 21-day online course that teaches you how to meditate. The site will also provides your own Stillness Sound that will act as an anchor during challenging times to take you back to a state of meditative calm.

**www.justgetflux.com** Flux creates a sunset on your screen to sync with the real sun setting outside, digital art imitating nature!

**MeditateApp** Allows you to select meditations for particular days and times. Meditations can be timed and saved in one of three modes: Meditation, Meditation with Affirmation, and Fall Asleep mode, which fades out the sound of the meditation gradually.

**www.mindfullnessdc.org** Mindfulness Bell is a simple reminder to pause and refocus attention. The bell, acts as a reminder to still the mind and be more present and can be set to ring at regular intervals or at random times.

**www.donothingfor2minutes.com** A full-screen video of waves crashing at sunset with a timer counting down for 2 minutes, along with the instructions, to provide you with a mini meditation break.

**www.stereopublic.net** Stereopublic is an app that crowd sources the quiet in the city. It provides a guide with 30-second video clips for those who crave a retreat from crowds.

**Lift** (iOS) Lift is a life coach app that helps you put your goals into action by providing positive reinforcement to create successful habits.

**AllTrails** This app is the essential trail guide that identifies local nature trails and walks.

**www.projectwildthing.com** Project Wild Thing is a movement to get kids and grown-ups playing outside more, roaming freely and reconnecting with the natural world.

# Acknowledgements

I would like to thank Jacqui Marson, chartered psychologist, whose infectious positive energy lights up a room, for filling the psychobubbles throughout the book with her wisdom and insights; Lisa Sanfilippo, yoga guru extraordinaire, for her unique "yoga snacks"; Dr Barbara Mariposa for sharing her humane and soulful mastery of mindfulness techniques; Howard Cooper, Rapid Change Therapist, for his effective mind control programmes and for sharing his touching personal life experiences with us; psychotherapist Jacqueline Palmer whose compassion and empathic approach to life touches all who have the privilege of knowing her and who also introduced me to the words of talented poet and writer Hollie McNish; Karina and John Stewart, founders of Kamalaya, who created an extraordinary sanctuary space to be able to digitally detox, nestled amongst the exotic natural wonders of Koh Samui; Arianna Huffington for giving her time to be interviewed and for personally taking the time to write to me, a gesture that reinforces the special qualities that make her an icon; Lewis Lapham for allowing us a window into his unique and cerebral vision of our world expressed by him in his interview with such intellectual elegance.

I would like to extend my special thanks to my editor Lisa Dyer and Alison Moss and the team at Carlton Publishing Group, who believed in this book from its initial concept and guided it through to its final incarnation so seamlessly, while allowing me the privilege of also having input into the concept and design of the book.

I also want to thank everyone who encouraged me to write this book along the way and shared with me their digitally overloaded experiences and my friends who encouraged me and gave me the space and time to sit and write (while following the 90/10 rule!), Linda Freeman for being an inadvertent catalyst for this book, Linda Malisani for her unfailing friendship, Jean Noel and Caroline for their loyal support, dear friends Kathleen and Edward who know the importance of real interaction and whose warm loving home is one of my happy places. Last, but not least, to my parents, thank you for your unconditional love and support and for always believing in me, and above all I want to thank my son Sash who has my heart and will always be my home.

# Index